LIFE BEFORE MAN

THE EARTH, ITS WONDERS, ITS SECRETS

LIFE BEFORE MAN

Reader's Digest

PUBLISHED BY

THE READER'S DIGEST ASSOCIATION LIMITED

LONDON NEW YORK MONTREAL SYDNEY CAPE TOWN

LIFE BEFORE MAN
Edited and designed by Toucan Books Limited
Written by Douglas Palmer
Edited by Mandie Rickaby and Robert Sackville West
Designed by Colin Woodman, Bradbury and Williams
Picture research by Adrian Bentley

FOR THE READER'S DIGEST, U.K.
Series Editor Christine Noble
Editorial Assistant Alison Candlin
Editorial Director Cortina Butler
Art Director Nicholas Clark

FOR THE READER'S DIGEST, U.S.
Editor Fred DuBose
Art Editor Eleanor Kostyk
Group Editorial Director, Nature Linda Ball
Production Supervisor Mike Gallo

Library of Congress Cataloging-in-Publication Data
Life before man.
          p. cm.–(The earth, its wonders, its secrets)
     Includes index.
     ISBN 0-7621-0138-5 (hardcover)
     1. Fossils, 2. Paleontology. I. Reader's Digest Association,
II. Series
QE711.2.L54 1999
560–dc21   98-48673
                                                              CIP

Printed in the United States of America 1998

FRONT COVER *Fossil ferns from the Triassic era, found in the Karnische
Alps in Germany. Inset: The slow moving* Triceratops *used its horns to
protect itself against predatory carnivorous dinosaurs.*

PAGE 3 Tyranosaurus rex *was the largest carnivorous dinosaur.*

CONTENTS

THE TROUBLING HISTORY OF LIFE ON EARTH

Few subjects have caused as much controversy as the evolution of life before man, with its profound challenge to traditional and religious ideology. Fossils supply the proof of ancient life and hold the key to science's greatest puzzle.

Our knowledge of past life on Earth can be compared with that of the frontiers of space or the deep oceans. Fossils provide the evidence to document the history of life, but the 'fossil record', as it is commonly known, is one area of science that has still not been fully investigated. This is largely uncharted territory.

Perhaps it is this sense of the unknown that provides the allure for those who study fossils – the palaeontologists – whether they are amateurs or professionals. Every year or so there are major discoveries, such as the dwarf fossil mammoths of the Arctic, the eight-toed tetrapods of Greenland or the baby dinosaurs still in their eggs. Only a tiny sample of the organisms of the past have been found as fossils and the further back in time the poorer the representation.

For palaeontologists today there is still a very real chance of finding fossil evidence of groups of extinct organisms that are completely new to science. Indeed, these types of find are constantly being made by both amateurs and professionals worldwide.

Initially, the history of modern palaeontological investigation focused on Europe and then North America, which meant that there remained vast continental areas which were

SIDETRACKED The Andes reveal the crisscross of footprints made by a Cretaceous dinosaur on what was once shoreline sandstone.

CLOSE TO THE BONE
Professional palaeontologists
extract dinosaur fossils in
Canada's Alberta Dinosaur
Provincial Park.

largely unexplored for fossils. Consequently, the major new discoveries in this century tend to have been made in the far-flung corners of Africa, South America, India, Asia, Australia and Antarctica. Some of these are still growth areas with huge potential for palaeontological field investigation, but this does not mean that the better known sites have been 'picked clean'.

Within the last few years, important discoveries of new fossil plants and animals have been found in one of the most investigated areas of geology in the world: the Welsh Borderlands. What has happened is that there is now a better understanding of how life forms can be fossilised, so that it is possible to prospect in rocks that previously would have been regarded as barren.

The real possibility of making a fossil 'lucky strike' is one of the main factors that ensures the continuing popularity of palaeontology. And anyone with a degree of persistence and willingness to learn has a good chance of making an interesting discovery. The significance of new finds may not be evident to the amateur immediately,

but professional help and guidance are available from museums and numerous palaeontological organisations worldwide.

THE DEVIL'S WORK?

Today, the existence of fossils as evidence of prehistoric life is taken for granted, but 400 years ago this was a matter of serious debate. Some believed that they were merely 'sports of nature' or 'works of the devil' made to fool gullible humans or even that they were inorganic artefacts growing in the ground like crystals, some of which just happened to resemble living creatures. Now the resolution to this great intellectual debate is known, but just how the fossil record came to be accepted for what it is forms the fascinating and often acrimonious history of palaeontology.

Over 300 years of collecting the fossil evidence, studying and arguing about it, has been required in order to arrive at any sort of consensus view. Now fossils are taken to include an enormous range of evidence of the existence of past life, from

MINERAL MOSS *The branch-*
like formation of minerals,
like this manganese
dioxide, has fooled naturalists
into regarding them as plants.

fossil DNA (deoxyribonucleic acid) to 3.5-million-year-old bacterial surfaces, and common shells and leaves to rare entire animal bodies of all sizes. The fossil record consists of all those remains of organisms from the past that have been entombed in sediments and rocks. Unfortunately, most of this record is still unavailable because it is covered by soil and vegetation or concrete, buried deep within the Earth's crust or has been destroyed by geological processes in the past.

DISCOVERING THE PAST

Human discovery of fossils dates back at least 80 000 years. Archaeologists have even found fossil shells in prehistoric burials of early (Neanderthal) humans. Ancient Greeks such as Pythagoras (6th century BC), Xenophanes (*c.*570-480 BC) and Herodotus (*c.*485-425 BC) are reported by later authors to have understood the true nature of fossils and to have realised that marine fossil shells found buried in rocks inland were the remains of organisms that once lived in the sea.

It was not until the Renaissance that significant advances were made in the understanding of fossils, when a few scholars were sufficiently independent of the prevailing medieval world view to trust their own observations. Leonardo da Vinci (1452-1519) was an acute observer of geological phenomena and his findings were a precursor to the modern science of palaeontology. Leonardo and later Italian scholars, such as Fabio Colonna, collected fossils and correctly related them to their

Verlag von J. F. Schreiber in Eßlingen bei Stuttgart.

POP AND ROCK *Showing early popular appeal, fossils are portrayed in this 19th-century German book as fascinating relics of the Earth's deep past.*

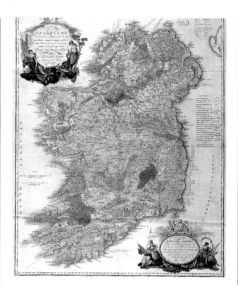

ON THE MAP *A geological map of Ireland in 1793 showed rock distribution, such as the Antrim lavas, without a sense of structural relationship.*

modern counterparts, many of which were to be found living in the nearby Mediterranean Sea. These were illustrated in some of the first treatises on fossils (such as Colonna's *De Glossopteris* of 1616).

The more questioning spirit of the 18th century's Age of Enlightenment encouraged the beginning of modern scientific methods and major developments in understanding the natural world. The French scientist Baron Georges Cuvier (1769-1832), for example, correlated the relationship between different rock strata and animal fossils in the Paris Basin. He also founded the science of comparative anatomy, which provided the basis for interpreting the fossil remains of extinct organisms, especially vertebrates.

In Britain, a self-educated engineer and surveyor, William Smith (1769-1839) realised that layers, or strata, of sedimentary rock occur in regular sequences, and that they can be traced over considerable distances. From newly dug canal sections, he was also able to demonstrate that particular types of fossil occurred in different strata. Thus the fundamentally important use of fossils for relatively dating rock strata was demonstrated to have great practical potential. For the first time, it became possible to make a geological map of a large area that included vertical sections showing the underground disposition and structure of the rocks.

William Smith completed a geological map of much of southern England in 1815, and this was followed by George Greenough's map of the geology of England and Wales in 1819. Richard Griffith published one of the first of the more sophisticated geological maps of an entire country, when he brought out his map of Ireland in 1834. Such maps have an important theoretical and predictive content, and realisation of this came at a crucial time in European history. The Industrial Revolution was under way and natural geological resources such as building stone, iron ore and coal were beginning to be exploited on a large scale. Landowners wanted to know whether there might be valuable mineral resources hidden beneath their fields and pastures.

During the 19th century, geological surveys were institutionalised as national, often quasi-military, organisations and much

palaeontological endeavour was tied to the study of rock strata and relating them to geological time as an aid to mapping. In North America there were still vast tracts of unexplored and unmapped territory as late as the middle of the century. The famous John Wesley Powell expeditions, from 1869 onwards, were the first to explore, date and map the Grand Canyon region, with all its magnificent landscape and geology.

The relative dating of all these rocks was dependent on identification of their contained fossils. Numerical or 'absolute' dating, which is taken for granted today, is based on the 20th-century discovery of the decay rate of radioisotopes of 'radioactive' elements contained in some minerals. In the 19th century, however, geologists just had a sense of the immensity of time that must have passed for the Earth's materials to be cycled to such an extent. They could see the sequence of erosion and removal of mountains; transport of eroded sediment to

FRENCH REVOLUTIONARY *Baron Georges Cuvier enlightened the 18th century by establishing the science of comparative anatomy.*

the sea; deposition and accumulation of the sediments in piles several miles thick; and, finally, the processes of mountain-building, intrusion and extrusion of molten igneous rocks, metamorphism and uplift, for the cycle to start all over again. There was an awareness that the Earth had been through several such cycles, so it was concluded that the geological history of the Earth had taken an immeasurable time. Since fossils were present for much of that deep time, it followed that life was equally ancient.

SCIENCE FICTION AS FACT

The discoveries of science in general have forced us to reconsider our view of the world and its formation within the Universe. The discovery of the history of life and how it has developed through time has led to one of the most traumatic reviews of the human position. The implication that life has an exceedingly long history ('without vestige of a beginning', according to James Hutton, one of the founders of modern geology) has many ramifications, which deeply troubled our Victorian forebears. That such concern and interest still prevail was amply demonstrated by the excitement over claims of possible primitive fossil life on Mars. The old 'egocentric' view that life is unique to Earth seems to have been dealt a fatal blow and, as a result, science

fiction is yet again becoming science fact.

The nature of the fossil record impinges on the human position in relation to other life – in no area more so than in human evolution and our demonstrable descent from ape-like ancestors. The study of fossils has shown that the modern version of Darwin's theory of evolution carries considerable weight, although it has taken a long time to become accepted generally.

THE TESTIMONY OF THE ROCKS

In the 19th century, the dominant Western viewpoint of the world was based on Judaeo-Christian ideology using a literal interpretation of the Old Testament. All creatures were the result of divine creation by God, it was believed, and as such their forms were fixed throughout existence. Many Christian scientists came to regard the fossil record not as a threat to belief but rather as an important source of divine truth, providing 'the testimony of the rocks' and revealing the wondrous work of God in all its diversity and detail. At that time, it was Darwin and Wallace's revolutionary theory of the origin of the species by means of natural selection that was to create such controversy in the understanding of life.

Ironically, despite Darwin's detailed investigations across a wide range of life forms and his

ALIEN CONCEPTS *Microscopic tubes within a meteorite from Mars have been claimed as fossil evidence for past life on the 'Red Planet'.*

awareness of the evidence of the history of past life, he did not involve fossils to any great extent as part of the proof of his theory. He knew only too well that the relatively primitive state of palaeontological evidence and understanding at that time raised more

REDISCOVERING AMERICA *Native Americans helped John Wesley Powell (on horseback, left) to 'discover' the geological and topographic wonders of the Grand Canyon in the late 1860s. In Wyoming, below, he discovered the world's largest accumulation of ancient lake deposits around the Green River.*

questions than it solved. Fossil evidence was too scanty and inconclusive to supply proof of any links between major groups of organisms, such as the reptiles and birds, now separated by independent evolution over a long time span.

The first important fossil evidence of a missing link between major groups was not discovered until 1861. Then, the remarkable *Archaeopteryx* fossil from Bavaria, with its mixture of reptilian and bird features, provided the first unequivocal evidence of an evolutionary tie-up between two major groups. But it was not until the end of the century that palaeontologists began a serious search for evolutionary lineages among the fast-growing museum stores of fossils gathered from all over the world. By 1874 the American palaeontologist Othneil C. Marsh (1831-99) was able to show that the fossil record of the horse family in North America demonstrated a more-or-less complete line of descent from the *Orohippus* of Eocene times through to the living genus, *Equus*.

MAKING SENSE OF IT ALL

Today, the study of living organisms has demonstrated that planet Earth supports a bewildering variety of life. From bacteria to sponges and bizarre fish in ocean depths, from insects to plants and snakes to cats on land, and from butterflies to birds in the air, life has found its way into almost all the nooks, crannies and niches that the surface of the Earth provides. Recently it has been

BONES TO PICK   O.C. Marsh (right) is almost as famous for his rivalry with fellow fossil hunter Edward Drinker Cope as for his own discoveries.

shown that viable and primitive life forms can survive within the surface rocks and sediments to depths of several hundred feet.

Scientists have attempted to catalogue and describe this array of life over the last couple of hundred years. And to try to make sense of it all, part of the endeavour has been to group like with like, to investigate and map the relationships between them. For example, is a bat more like a bird than a human because it has wings or does its hair and bearing of live young relate it more closely to man?

This system of classification – technically known as taxonomy – attempts to group all living organisms into a sort of hierarchy, based on similarities of structure and origin. The animal kingdom is the top of the hierarchical tree as far as man is concerned, for example, and this he shares with dogs, cats and all other animals with a backbone. Farther down that tree, man is still grouped with cats and dogs within the class division Mammalia, because all share mammalian characteristics of hairy bodies and the suckling of their young. Lower down still, humans and cats and dogs branch off in different directions. Humans are subdivided with monkeys and apes into the order Primates, whereas cats and dogs form part of the Carnivora. Thus the classifications carry on subdividing until the most fundamental unit of the hierarchical

FALSE COLOURS   More than 500 million years have turned the exoskeleton of the shrimp-like Jianfengia, *from China, a rusty red.*

tree is reached: the species. In this way, scientists can organise all creatures into discernible and specific groupings, which makes it possible to trace the biological affinity between them all.

Modern man – or *Homo sapiens*, to use the proper species name – has close relatives in a number of fossil and extinct species, such as *Homo erectus* and *Homo habilis*. The problem is that whereas all observations of *Homo sapiens* can be done at first hand – obviously – recognition of *Homo erectus* and *Homo habilis* as species has to be based on different criteria because they ceased to exist so long ago. Consequently, there is no direct knowledge of their real lives, and designation has to be based on a study of their preserved fossil anatomy.

Evidently, a fossil species is a very different entity to a living being, and there is always an interpretative element to its definition. Consequently, as with interpretation of the law, there is always the potential for disagreement, even – or perhaps especially – between experts. And this is never more pronounced than when it comes to human taxonomy because it impinges upon our view of ourselves and our fossil ape ancestry, which many still find unacceptable.

HIERARCHY OF LIFE

Despite the existence of a biological system of classification for all plant and animal life, the total number of species on Earth is far

LIFE AND TIMES

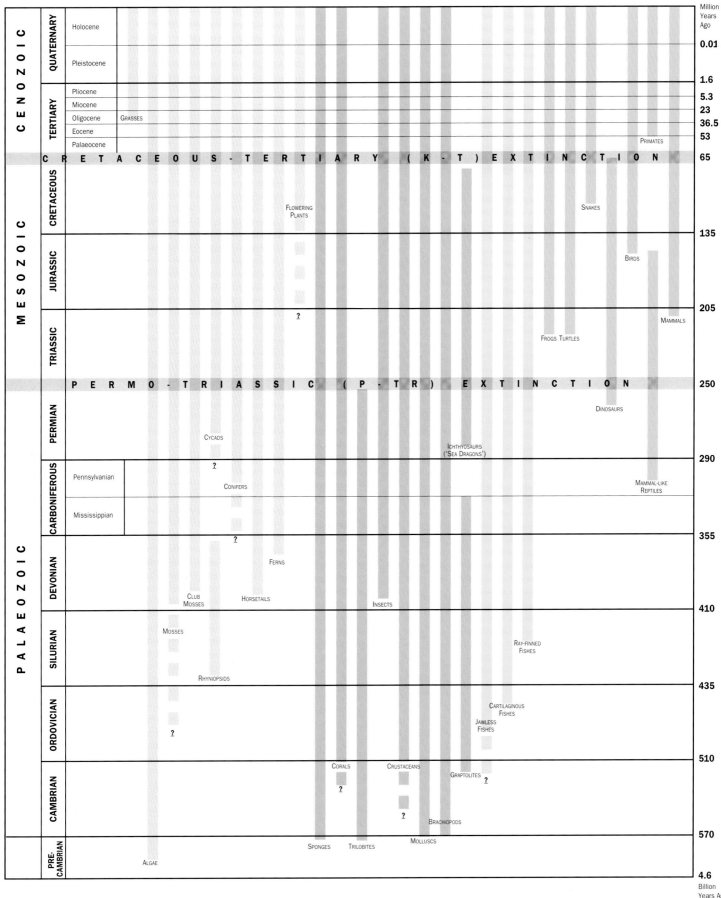

			Million Years Ago
CENOZOIC	QUATERNARY	Holocene	
			0.01
		Pleistocene	
			1.6
	TERTIARY	Pliocene	5.3
		Miocene	23
		Oligocene	36.5
		Eocene	53
		Palaeocene	65

CRETACEOUS-TERTIARY (K-T) EXTINCTION — 65

Grasses

Primates

MESOZOIC	CRETACEOUS		135
	JURASSIC		205
	TRIASSIC		250

Flowering Plants

Snakes

Birds

Mammals

?

Frogs Turtles

PERMO-TRIASSIC (P-TR) EXTINCTION — 250

PALAEOZOIC	PERMIAN		290

Dinosaurs

Cycads

Ichthyosaurs ('Sea Dragons')

	CARBONIFEROUS	Pennsylvanian	
		Mississippian	355

?

Conifers

Mammal-like Reptiles

	DEVONIAN		410

?

Ferns

Club Mosses

Horsetails

Insects

	SILURIAN		435

Mosses

Rhyniopsids

Ray-finned Fishes

	ORDOVICIAN		510

?

Cartilaginous Fishes

Jawless Fishes

	CAMBRIAN		570

Corals

Crustaceans

Graptolites ?

?

?

Brachiopods

	PRE-CAMBRIAN		4.6 Billion Years Ago

Sponges Trilobites

Molluscs

Algae

SLICE THROUGH TIME *The Colorado River has cut through 300 million years of Earth's history, from Permian sandstone to Cambrian shale.*

from clear. This is because the description of species has been conducted over 300 years by many thousands of naturalists, scattered all over the world, working and publishing in many different languages. At present, the estimate of known and described living species is between 1 and 1.75 million. But experts are well aware that this figure represents only a fraction of the total number of species that are alive today. For this, estimates vary hugely, from 3 million to a massive 30 million. One thing is sure: with the present rate at which species are being lost due to the destruction of their habitats (estimated at between 10 000 and 25 000 species per year), there are many organisms that we will never know about.

HEAD COUNTS FOR THE PAST

When it comes to fossil species, the record is inevitably more complicated because of the increased difficulty of definition. Estimates of the total number of described fossil species are variable at between

300 000 and 500 000 – clearly a tiny sample of everything that has existed throughout the 3.5 billion years of the history of life on Earth. To gain a better idea of the proportion of all life as represented by the fossil record requires some juggling of 'guesstimated' figures. First of all, one can make a conservative estimate that each species lasted 5 million years over the 550 million years of the Phanerozoic era (from the first evidence of abundant life to the present day). This means that there were approximately 100 turnovers of all animal and plant life during that time span. Secondly, by estimating that there were 10 million species alive at any one time this would give a conservative global total of 1000 million species. Taking the known fossil record at a slightly high estimate of 500 000 species, then it is apparent that this is indeed a pathetically small sample at about 0.05 per cent of that global total.

Such calculations depend on knowing the rate of species turnover through time and the degree of diversity within a species. In recent years, estimation of species duration has become of greater interest to scientists because of its importance in calculating any changes in the diversity of life forms and, more particularly, the present rates of extinction. The average life span of a species turns out to be between 1 and 10 million years, although many animals are way out of this range. Some bird and mammal species last only between 200 and 400 years. Estimates of the total number of species ever to have existed are, not surprisingly, the most variable figures of all. The range is between 50 and 4000 million, so any calculations of the representativeness of the fossil record are still

crude and contain large margins of error. But while the known fossil record is undoubtedly a poor representation of life when it comes down to individual species, the overall position is rather less gloomy when it comes to the broader groupings within the biological classification system. This means that whereas there may not be a very detailed fossil record of, say, the Australian dingo (species, *Canis dingo*), there may be sufficient evidence of other relatives – such as the jackal – to form a comprehensive picture of the Canidae family to which they all belong.

BIASED RECORDS

A crucial element in any interpretation of the fossil record is the fact that it can only provide evidence of the existence of preservable organisms, and so is highly biased. Generally, only those organisms that have hard or mineralised parts of their bodies – such as bones or shells – can be fossilised. There are many groups, especially of soft-bodied organisms, that are extremely unlikely to be preserved. What is remarkable is that, given ideal preservational circumstances, creatures as small and

unprotected as bacteria can be fossilised and survive many hundreds of millions of years and successive upheavals of the Earth to be found by palaeontologists today.

Even the rocks that contain fossils are biased in their own way, presenting a geological record in which only certain sedimentary layers have stood the test of time. Most preserved sedimentary rock consists of shallow sea deposits such as limestone, sandstone and shale. Terrestrial sediments – those accumulated on land – are not well preserved, generally, and deep-sea deposits are hardly preserved at all. Consequently, most fossil discoveries consist of the skeletons of marine organisms.

Palaeontologists know quite a lot about the various snails, clams, trilobites and ammonites that once lived in the shallow seas that flooded the margins of the ancient continents, but knowledge of what lived on the land and in the air is relatively poor by comparison.

THE FIRST SIGNS OF LIFE

Undoubtedly one of the greatest modern palaeontological breakthroughs has been the discovery of traces of life in rocks from

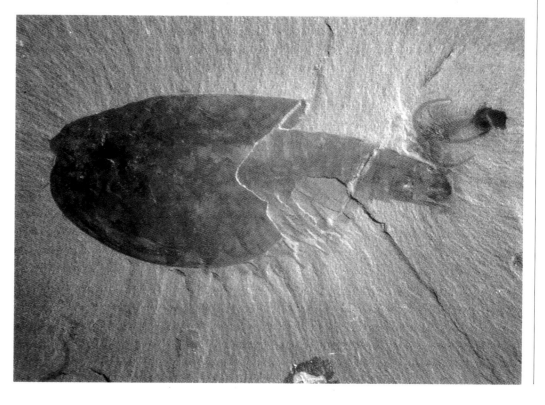

CAMBRIAN SQUASH *Despite being flattened to thin film, these fossil arthropods are remarkably well-preserved, with their appendages intact.*

PUZZLING FRONDS *Strange leaf-shaped organisms, preserved in Late Cambrian sandstone, may have been related to living sea pens.*

the Precambrian period, when no forms of 'obvious' life were apparent. These fossils of primitive life – microscopic organisms at the cellular level – have been found in

NOW AS THEN *Primitive single-celled bacteria still build up layers of sediment – known as 'stromatolite' mounds – on the coast of Western Australia.*

locations such as Swaziland and Western Australia, in Precambrian rocks varying in age from 600 million to 3.5 billion years old.

Having survived against all odds of preservational bias, these ancient organisms are the oldest forms of life ever to have been discovered. They take the form of 'stromatolitic' mounds: structures built up of layers of 'prokaryotic' – single-celled organisms without a nucleus – bacteria that have interacted with fine-grained sediments in shallow, warm seas.

Very little is understood about the crucial first steps in evolution, but it is likely that these primitive organisms were preceded by the long phase of chemical development in which the building blocks of life – the biomolecules – were built up. It is believed that the first viable single cells of true life must have evolved about 4 billion years ago. It isn't until about 2.3 billion years later that evidence of the first multicellular organisms appears, in the form of algal fronds. The first indications of multicelled animals do not appear until about 1 billion years ago. And well-organised multicelled (metazoan) animals, the Ediacarans – something like our modern jellyfish – are not apparent for another 400 million years. This means that it took at least 4 billion years for life to evolve even to this – still primitive – point.

The first appearance of organisms with skeletons 550 million years ago marks the end of Proterozoic time (which means 'first life') and the start of the Phanerozoic era (literally 'obvious life'). The first signs of these have been discovered in

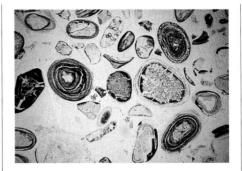

CIRCLES OF LIFE *A section of Precambrian sedimentary rock from Canada reveals the concentric growth of a number of filaments of algae.*

earliest Cambrian rocks and, while they consist predominantly of shells, scientists have been surprised at the extraordinary diversity of life that must have existed in those times. The famous Burgess Shale, high in the Canadian Rockies in British Columbia, has provided a remarkable window on life in the Cambrian seas. Just one single location has provided more than 11 000 fossil specimens, most of which were completely new to science at the time of their discovery in 1910.

THE ULTIMATE JIGSAW

Every year, new finds are made which throw a whole new light on the history of life before – and after – man. Nevertheless, palaeontologists have got a long way to go to catch up with their biological colleagues, who have a much better idea of the totality of life today. For many biologists, the fossil record is an endless source of frustration because it is like an enormous jigsaw puzzle for which there are only a few pieces available. Furthermore, the total number of pieces is unknown, as is the final picture. Some of the pieces do seem to fit together but they only give little glimpses of parts of the overall scene.

Just 30 years ago, the eminent biologist and Nobel prize winner, the late Sir Peter Medawar, dismissively remarked that all that remained for palaeontology to do was to 'fill in the details of the parish register of life in the past'. How wrong he was.

IN THE BEGINNING 1

CROWD PULLERS *A host of empty ammonite shells fill this Jurassic seabed surface.*

LIKE ANY GOOD STORY, THE HISTORY OF LIFE BEFORE MAN HAS A BEGINNING, A MIDDLE AND AN END. AND THESE THREE BROAD PHASES OF DEVELOPMENT TRACE THE EVOLUTION OF LIFE FROM THE FIRST MARINE ORGANISMS TO THE ANCESTORS OF EARLY MAN. MOST FORMS OF MARINE LIFE MANAGED TO EVOLVE WITHIN THE FIRST PHASE BUT, OUT OF THE WATER, THE PICTURE WAS QUITE DIFFERENT — AND WOULD BE TOTALLY UNFAMILIAR IN TODAY'S TERMS. IF WE COULD GAIN A GLIMPSE OF SILURIAN LIFE, THE WINDOW WOULD OPEN ONTO THE GREENING OF THE EARTH WHEN FIRST PLANTS AND THEN ANIMALS LEFT WATER FOR DRY LAND. THIS WAS TO PROVE ONE OF THE GREATEST REVOLUTIONS IN THE HISTORY OF LIFE.

ROCK WREATH *Fossilised sea lilies from Carboniferous mud.*

BESIDE THE SILURIAN SEA

The stone-grey fossils in the museums of today are the only record we have of the early history of life on Earth. Yet recent scientific research has brought the multi-coloured creatures that once inhabited our planet to life.

From about 400 million years ago, during the Silurian period, many forms of life which had originally evolved in the sea made their first tentative movements towards the land. So it was, at the point where land and sea met, that flora and fauna struggled to emerge onto the shore. The ancient shoreline looked much the same as modern beaches. The rocks, sand and mud were no different: waves and surf pounded the beach; tides rose and fell; and rocks were slowly whittled away by the forces of erosion and then deposited as sand in a process that has continued inexorably, year by year, aeon by aeon. Even the stranded jellyfish and the seaweed looked the same: strands of marine algae in shades of brown, green, red and yellow lay scattered about, torn from their rocks and sand anchors. Then, as now, objects were blown in from

LONE SURVIVOR *The chambered shell (right) of the present-day* Nautilus *(above) marks it out as the remaining descendant of the ancient cephalopods. Like its fossil ancestors, the* Nautilus *uses its shell to maintain buoyancy at a chosen depth.*

the ocean, especially after a storm, or stirred up and excavated from the sea floor before being washed up and stranded by the retreating waves at the edge of the highest tidemark. Tangled in amongst the weed were dead sea creatures and whole shells of various kinds, some of them very fresh. Other shells were broken after being tumbled about by the waves.

Farther down the beach, at low tide, the sand flats looked, at first glance, like a wet desert, devoid of life, and consisting of nothing more than sand ripples and patches of shell debris. But there was plenty of evidence of life all around. The surface of the Silurian beach was broken by wormcasts and crisscross tracks left by creatures which had scuttled into the sand and mud.

Some of the shells would have been vaguely familiar to a beachcomber today, but many were quite unrecognisable. In one crucial respect, however, the Silurian beach was very different from today's. It was more

BEACH BATTLEGROUND *Aeons ago, life struggled to emerge from the sea and gain a foothold on land, on shorelines that resembled our own (left).*

representative of the totality of life over 400 million years ago than a modern beach is representative of life in the present. The world's fresh waters had not yet been invaded by animals, and plants had only just begun to take root in the landscape. As a result, life in and beside the Silurian seas was the predominant form of life on Earth. And the shore was the best place to view it.

THE EARLY NAUTILUS

Poking out from amongst the weed of the Silurian shore, the most obvious objects were strange, straight, cone-shaped shells,

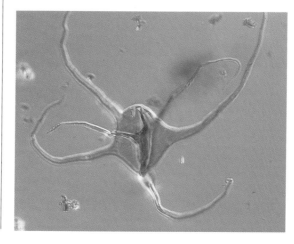

about 1 ft (30 cm) long and 1½ in (3.8 cm) in diameter, some with colour stripes and others with zigzag markings. The empty shells were remarkably thin and light, but were strengthened by a series of curved internal walls that divided the length of the cone into chambers. A neat hole perforated the wall of the last chamber, extending as a tube back through the entire length of the cone; inside, the surfaces of the shell had a pearly sheen that would have been familiar to us today from molluscs, such as clams and sea snails, found on modern beaches.

The creature occupying the cone shell looked very much like a small squid, with a soft 'slimy' body, well-developed eyes and tentacles. If placed in a pool, it would have tried to dart away, shooting backwards like a rocket by using water-jet propulsion. When it was not under threat, *continued on page 20*

SILURIAN SEA FOOD *Microplanktonic organisms, like this fossil acritarch, have long formed the basis of the marine food chain.*

THE SILURIAN SEA

At first, an underwater Silurian scene would seem to be similar to anything found in warm, sub-tropical waters today. There were plenty of shellfish, coral and fish around, but a closer look would reveal that many of the creatures were in fact quite different.

Most common were filter feeders, which sieved microorganisms and organic detritus from the surrounding water. Many lived permanently attached to the seabed and grew mineralised skeletons for protection and support. Corals formed reefs, which sheltered sponges, moss animals (bryozoans), sea lilies (crinoids), bivalved brachiopods and clams. The crinoids and many brachiopods grew permanent attachments to the seabed. They are regularly found as fossils in ancient rocks.

Most mobile animals, such as snails, were restricted to the seabed. They, like their modern counterparts, were algal grazers. Starfish and trilobites were scavengers and some could burrow. A few trilobites could swim, probably to escape from predators such as the eurypterids. These arthropods were the biggest animals of the time, some growing to 8 ft (2.4 m) in length. Most were scavengers but some were active hunters of small animals. The distinctive trilobites and eurypterids are now extinct, as are the bizarre carpoids.

There were active swimmers in these ancient seas, such as fish and squid-like cephalopods. Most fish were jawless and toothless, and fed on tiny organisms or organic debris (although the first jawed fish with teeth had evolved by this time). The most abundant and active hunters were the cephalopods. With streamlined, conical shells, these ancestors of the surviving pearly *Nautilus* swam by water-jet propulsion and caught prey with their tentacles.

TRILOBITE

CHAIN CORAL

GASTROPOD

EURYPTERID

CRINOIDS

CEPHALOPOD

BRACHIOPODS

JASMOYTIUS

BRYOZOAN

CALCICHORDATE

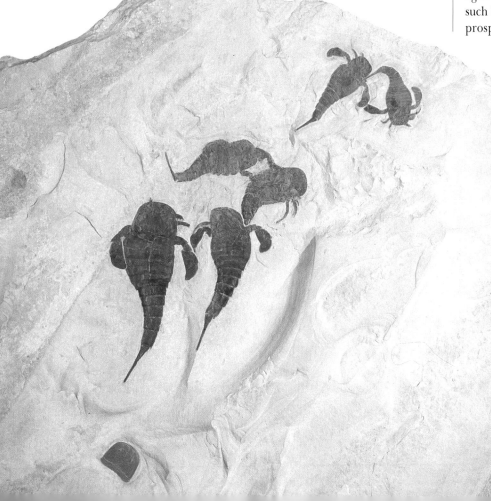

WATER SCORPIONS
Eurypterids, with swimming paddles and jointed exoskeletons (fossils below), were among the first animals to move from the seas into fresh waters (above). Although now extinct, they are distantly related to modern scorpions.

however, it swam more slowly head first, crawling over the rocks and seabed with its tentacles in search of food. It stayed buoyant by controlling the proportion of liquid to gas in its chambers with the extended tube called a siphuncle. A soft, slimy body and a pearly (nacreous) shell are among the distinguishing features of molluscs today. The tentacles and eyes are characteristic of cephalopod molluscs, such as living squid, octopus and cuttlefish. The final clue – the chambered shell – identifies this Silurian creature as a nautiloid cephalopod, which still has one surviving close relative, the pearly *Nautilus*, an inhabitant of the Indian Ocean.

The shells of nautiloid cephalopods were a common sight on the Silurian shore, coming in different shapes (some of them were coiled), colours and sizes. Nautiloids were amongst the largest animals in these ancient seas. Although most of the cones were only a foot or so long, the fossil remains of some rare nautiloids indicate lengths of more than 6 ft (1.8 m). The only other creatures that competed for size were fearsome-looking animals called eurypterids, which are now extinct.

ANCIENT ARTHROPODS

Fossil eurypterids have been found in western New York State, nearly 8 ft (2.4 m) long and armed with lobster-like claws; others had tail spines as well. Such creatures looked like, and were distantly related to, the terrestrial scorpions of today, and as a result have been called sea scorpions.

Eurypterids were not just the largest, but also the best-armed hunters of the time. A tough jointed 'shell', or exoskeleton, covered their bodies, providing protection against the elements. The possession of such body armour was to further the future prospects of other arthropods, including insects, crustaceans and millipedes, although it did not save the eurypterids themselves from becoming extinct at the end of the Permian period, some 250 million years ago.

Their armour posed problems as well as conferring advantages. As with the medieval knight, the armour, which consisted of individual plates, had to be jointed to allow movement. But to enable the creature to grow, the whole suit had to be discarded and a larger one built from time to time. Eurypterids shed their old suits simply by splitting them along the seams between the plates. Inevitably, this process of moulting, or ecdysis, created periods when the animals were very vulnerable to attack (in much the same way as modern soft-shell crabs are). Fortunately for

Overall, the trilobites were very successful animals, and their remains – usually in the form of individual plates of moulted body armour – would have been as common on a Silurian beach as the remains of crabs on the beaches of today.

SILURIAN MOLLUSCS

Amongst the 'litter' of the Silurian beach, there were also some more familiar-looking shells. Coiled sea snails (gastropods from the same class of molluscs that now includes cowries, limpets and slugs) came in a range of colours, shapes and sizes, along with five or six different kinds of clams – just as we might find on a beach today. Each type of animal inhabited its own particular niche, so as to avoid too much competition for food resources. The snails, for example, were plant eaters, living on – and off – the seaweed.

As with the bivalve molluscs of today (for example, cockles, mussels, oysters and scallops), the Silurian clam shells ranged from fat to thin, and from smooth to knobbly, although they were distinctly smaller than modern clams. The soft body of the clam was completely enclosed and protected

palaeontologists, however, the repeated moulting has increased the number of potential fossil remains of the animals.

Pairs of jointed legs, and the possession of an exoskeleton which has to be moulted from time to time, are characteristics of the arthropods – the phylum that includes a range of invertebrates from insects to crustaceans. Among the most common ancient arthropods were the trilobites, also now extinct. Most were only 2 or 3 in (5-7.5 cm) long, and some would probably have been found on the Silurian beach, rolled up and looking like knobbly golf balls. In general

appearance, trilobites were a little like today's horseshoe crabs (limulids), but they had more distinct heads, with insect-like, multi-lensed eyes, tail shields and flexible bodies.

The trilobites were scavengers, rather than fearsome predators, probably feeding on plant or animal detritus. Some of them may have removed organic material from the sediment, as they ploughed or burrowed their way through the mud and sand on the sea floor. Others, with small, light bodies, may have been able to swim and capture swimming microorganisms for food as they did so.

by two hard, symmetrical shells, joined by a hinge. Living inside a hinged box like this has enabled the clam to close its doors on the outside world, providing a degree of safety from predators.

Some of the shells on the Silurian beach

DATING OUR ANCESTORS

Over a century ago, Charles Doolittle Walcott, one of America's most brilliant palaeontologists, discovered pieces of bone in an Ordovician sandstone deposit, some 500 million years old, near Canon City, 30 miles (48 km) south-west of Colorado Springs. Since bone is characteristically a body tissue of vertebrates, his discovery was unequivocal evidence that vertebrates had evolved by early Ordovician times.

were small, smooth or finely ridged, and looked remarkably like the 'nut' shells, or nuculids, of today. 'Nut' shells are distinguished by their many teeth and sockets, which act as an articulating hinge between the two valves. This group of clams has a remarkably long history, stretching back over 470 million years into the Ordovician period. The animals burrowed – and still do today – in sandy shallows, 'hoovering' up the surrounding mud as a source of food. This particular niche has clearly provided them with a continuing safe haven, without too much competition, and they have had no reason to move or change much over time. In evolutionary terms, they are a good example of a stable and conservative group of animals – of a type that is often referred to as the 'living fossils'.

Although, at first glance, they might have looked a little like clams, most of the bivalve shells on the Silurian beach belonged, not to molluscs, but to another group of shellfish, the brachiopods, that are rare today, except in New Zealand and Japan. Most of them resemble Roman oil lamps, with one shell bigger than the other and a

hole perforating the point of its 'beak'. The name, meaning 'arm-footed', is something of a misnomer, however, for it applies to the looped, feathery structure within the shell (*brachion* is Greek for 'arm'), which was originally thought to be for movement (*pous* is Greek for 'foot'). We now know that it is used for feeding and breathing. The tiny tentacles set up a flow of water through the shell, sieving food particles and extracting oxygen from the passing water.

SHELL LIFE FOR BRACHIOPODS

The lifestyle of the brachiopod is basically the same as that of most bivalve molluscs. In essence, they are both small filter-pump mechanisms that live on the seabed. One major difference, however, is that most brachiopods lack mobility. They start life anchored to the sediment or to some object on the seabed by a fleshy stalk that extends through the hole in the shell beak. This means that they cannot burrow into the sediment for protection. In evolutionary terms, this does not seem to have mattered much to begin with, and in the Palaeozoic era, brachiopods were clearly

more successful than bivalve molluscs. But it is likely that, with the development of the evolutionary 'arms race' between carnivorous predators and herbivores or filter-feeders, the only animals to prosper in the long run were the mobile ones, like the bivalve molluscs; which is why there are many more molluscs around today than brachiopods. They were the ones that could escape the predators by burrowing into the sediment of the seabed – at least, until the predators took to burrowing after them.

Despite these limitations, brachiopods evolved a variety of shell forms, reflecting (like the bivalve molluscs) differences in their way of life and where they lived. In turbulent underwater environments, especially in shallow waters constantly buffeted by waves, shells had to be physically tough, thickened or folded into pleats or ribs, as in the case of the rhynchonellid brachiopods and the cardiolid bivalves in the Silurian sea.

In shifting sand and soft mud, those shellfish which did not burrow had to develop a different type of shell to allow them to stay on the surface of the seabed. One strategy – known as the 'snow shoe' solution – was to maximise the surface area of the shell relative to its overall weight. This produced a flat shell form (like the now-extinct winged bivalve mollusc, *Pteritonella*), sometimes enhanced by marginal spines (as in the case of chonetid brachiopods, *Protochonetes*, in the Silurian era). However, this reduced the internal space drastically and limited the overall size to which the creature could grow.

An alternative strategy was the hemispherical form, in which the lower valve was cup-shaped and covered with spines for support, and the upper valve was flat (as with strophomenid brachiopods in the Silurian era). The ideal form,

SEA SNAIL *The fossilised shell of a 425 million-year-old mollusc,* Poleumita, *looks well enough preserved to have come from a modern beach.*

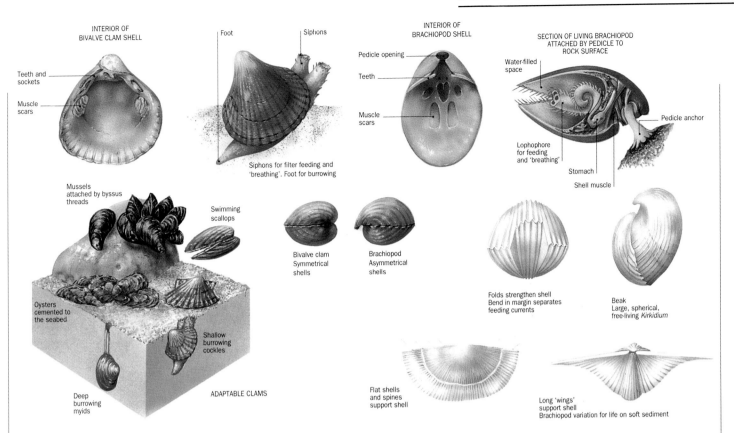

INTERIOR OF BIVALVE CLAM SHELL

Teeth and sockets

Muscle scars

Foot

Siphons

Siphons for filter feeding and 'breathing'. Foot for burrowing

INTERIOR OF BRACHIOPOD SHELL

Pedicle opening

Teeth

Muscle scars

SECTION OF LIVING BRACHIOPOD ATTACHED BY PEDICLE TO ROCK SURFACE

Water-filled space

Pedicle anchor

Lophophore for feeding and 'breathing'

Stomach

Shell muscle

Mussels attached by byssus threads

Swimming scallops

Oysters cemented to the seabed

Shallow burrowing cockles

Deep burrowing myids

ADAPTABLE CLAMS

Bivalve clam Symmetrical shells

Brachiopod Asymmetrical shells

Folds strengthen shell Bend in margin separates feeding currents

Beak Large, spherical, free-living *Kirkidium*

Flat shells and spines support shell

Long 'wings' support shell Brachiopod variation for life on soft sediment

which maximised the size of the filter organs and hence the rate of pumping water and food consumption, was produced by the large, almost spherical, shells of the pentamerid brachiopods and the cardiolid bivalves in the Silurian era. Even if the bottom of the shell sank into the sediment, the valves opened above the surface.

Other variants included the one major group of brachiopods, the lingulids, able to burrow. Evolving in the Ordovician era and surviving until the present – over 460 million years – without much change in their lifestyle or shell form, the lingulids are another example of 'living fossils'.

SPINY-SKINNED ECHINODERMS

The echinoderms, whose name means 'spiny-skinned', are familiar enough today – in the form of starfish and sea urchins. Yet they were also to be found on the Silurian beach; indeed, their fossil record stretches right back into the Ordovician era.

The 'shells' of the echinoderms were different from those of the brachiopods or the molluscs which, when closed, clammed up the creature inside. Instead, echinoderms consisted of articulated plates made from a chalky mineral, called calcite, and arranged symmetrically. Overlaying the

shell was a thin skin of tissue covering the base of the spines on the outside surface. Muscles within the tissue enabled the creature to move these spines, as with sea urchins today. Modern starfish have softer shells than sea urchins, with the result that the 'skin' is more apparent: a starfish skeleton found on a beach of today often consists of nothing more than a series of hard needles, or spicules, embedded within the tissue. But this was not always the case: the 'shell' of many Palaeozoic starfish was made up of much stiffer plates, and this has preserved them more readily as fossils.

Sea urchin shells last some time after death because the calcareous plates hold together well, even after the spine muscles have rotted and the spines themselves have fallen off. Empty sea urchin shells are not uncommon on our beaches today, but most of them are fragile and break easily under the impact of the pounding surf or grinding shingle. To have been preserved intact as

SHELL SHAPE *In the Silurian era, brachiopod shells, like Doleorthis (below) were more common than bivalve molluscan clams (above), with their different shell symmetry. Today, the more adaptable clams have taken over.*

fossils, they had to be buried fairly quickly in the sediment.

Protected by the articulated plates that are characteristic of echinoderms, the sea lilies, or crinoids, were close relatives of the starfish and sea urchins in Silurian times. Then, as now, these curious animals looked plant-like in a stiff sort of way. Rooted in the sediment or to a rock, they had upright flexible stems, on top of which there was a cup fringed with feathery arms. Sea lilies were filter feeders, fanning out their arms in the water to act as a web or sieve, and then trapping microscopic organisms as they drifted by. They avoided competition with all the other filter-feeders living on the ancient seabed, such as the brachiopods and bivalve molluscs, by evolving stalks, rather in the way that plants grow higher and higher to get more light.

The Silurian seas were also home to a curious group of now-extinct creatures, known as graptolites because, when fossilised, their organic skeletons became

ANCIENT AND MODERN
The sea urchin's protective spines (right) are mounted on tubercles, which are apparent in the fossil urchin (below).

carbonised with the heat and pressure and began to look like graphite pencil lines on the rock surface. In Silurian times, however, their stems, a tenth of an inch (2.5 mm) or so in length, would have looked like some geometrically patterned plant, stranded by the receding waves in clusters on the shoreline or entangled with the algae farther up.

ANIMAL, VEGETABLE OR MINERAL?

For over 150 million years, from the Middle Cambrian (about 535 million years ago) to the Lower Devonian (about 395 million years ago), the seas and oceans of the world teemed with graptolites, rather like jellyfish today. In the early 18th century, scientists misinterpreted their fossils and lumped them together with a range of flat, moss-like markings in the rocks, some of which were not fossils of organisms at all but were, in fact, mineral growths. Since then, graptolites have been considered variously as plants, worms, cephalopod molluscs and hydroid coelenterates, distantly related to corals and jellyfish. It was not until the 1930s, after a meticulous study of their skeletal structure, that the Polish palaeobiologist, Roman Kozlowski, resolved the zoological affinities of the graptolites, pinning them down – despite their apparent plant-like form – as

FIVE-FOLD SYMMETRY
An impression on a rock surface produces the readily recognisable petal shape of the fossil starfish, Platanaster.

advanced invertebrate hemichordates. The animals consisted of branched and overlapping tubes, arranged like fretsaw blades, forming a series of serrated stems and filled with a necklace-like colony of tiny gelatinous zooids. The tube walls themselves were made of a tough but flexible material, which was secreted and built up in layers by the zooids. Free-floating colonies of the creatures were carried by ocean currents around the world, ranging in appearance from many-branched shrubby forms to 'matchstick' single branches.

THE FIRST FISH

Because of ever-vigilant scavengers, it is unusual to find whole dead fish on modern beaches, although individual bones are common amongst the shell hash. In Silurian times, however, there were far fewer scavengers, and therefore a better chance of finding something interesting. Nevertheless, the problem with looking for Silurian fish – and one that bedevils much scientific endeavour – is knowing what to look for, when you are searching among fragments and remains that have been transformed by fossilisation into something very different from the living form.

Today we think of a fish skeleton as basically a backbone with lots of long thin ribs and spines running off it. But this view is based on the dominant living fish, which are bony, or osteichthyan, fish such as salmon, trout and cod, with a skin of thin, overlapping scales and, for the most part, a flattened body. But there is another type of modern fish – the cartilaginous, or chondrichthyan, fishes, represented today by the sharks and the rays.

Typical leftovers on a Silurian beach might have included rather leathery plates, with one surface covered with random knobbly patterns. There might also have been patches of leathery skin – with the texture of sandpaper or, rather, of shagreen (sharkskin), smooth in one direction and rasping in the other. When the skin disintegrated further, it broke down into individual, pointed scales,

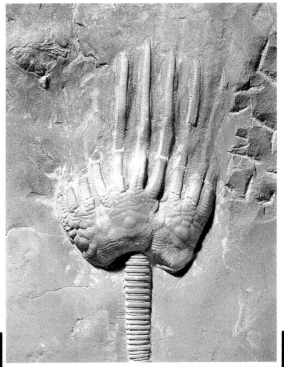

SEA LILIES *The fossil crinoid,* **Marsupiocrinus** *(right) is preserved with anchoring stem, body cup and feeding arms. A living feather star (below) demonstrates how crinoids fed with their arms outstretched to catch organic detritus in the current.*

a millimetre or so in size, which have been preserved in the fossil record.

The internal backbone, or vertebral column, of these fish did not fossilise well because it was made of cartilage-like material, similar to that found in sharks today. But Ordovician skin plate fragments have survived and been identified as belonging to *Astraspis*, a jawless (agnathan) fish.

One later group of agnathans – the cephalaspids – had a horseshoe-shaped head with backward-projecting spines at either end, closely spaced small eyes on top, and a small oval mouth on the flat lower surface. The tail had a more familiar 'fishy' appearance with a scaly surface and a fin.

ARMOURED FISH *Many Silurian fish were ungainly creatures – jawless, toothless, and with their heads covered in a leathery armour.*

LIFE ON THE EDGE *The stems of graptolites had serrated edges that once housed colonies of tiny creatures, called zooids.*

The tough head shield of the cephalaspids was made of a semi-rigid 'leathery' material that was closer to true bone and covered with a thin layer of skin, like a sea urchin skeleton.

Another group of agnathans, called the thelodonts, were more fish-shaped and covered in a shark-like shagreen of overlapping, backward-pointing scales. The mouth was right at the front of the animal and, like that of the cephalaspids, lacked supporting jaws or teeth. These agnathans were very different in their feeding habits from modern bony fish with their small sharp teeth.

Agnathans could have eaten only very soft organic material, such as mud or decaying plants for example. Their jawlessness – a feature today of lampreys and hagfish – was a sign of their primitive state. Jawless fish were the earliest fish of all.

THE FIRST VERTEBRATES

Tiny eel-like creatures, only an inch or so long, flitted across the rock pools of the Silurian shore. They would have been quite hard to see at first because their bodies were translucent, and they were constantly on the move, swimming about with a sideways wriggling motion.

Through the translucent flesh of their body, it would have been possible to see a series of diagonal muscles, running from head to tail. Within these, along the upper part of the back, ran a faint tube or rod which gave a slight stiffness to the body. The rod was very flexible, so that when the muscles on either side contracted, it bent easily into a sinuous series of S-shaped curves. These body waves moved from the tip of the tail forwards, propelling the creature through the water.

The head was quite peculiar. Below the eyes, there was a pouch filled with a framework of tiny and very spiky, cone-shaped teeth – or conodonts – from which the creatures got their name. The teeth pointed inwards and were arranged in a short tube-like apparatus, which shot forwards through the mouth and opened up to grab the prey in its mesh of spines. As the apparatus returned into the mouth, the meshwork closed tighter, and some blade-like teeth chopped the prey to pieces.

This description of the conodont is based on recent fossil discoveries, which have led to the conclusion that conodonts were true vertebrates, albeit at a very early stage in evolution. Their fossil record extends back to the Late Cambrian era (some 515 million years ago) and possibly a further 65 million years to the Early Cambrian era – which would make them the first vertebrates.

Fossil evidence suggests that the first fish with jaws had evolved by Silurian times. These were a group of small fish, just a few inches long (up to 20 cm), called the acanthodians, or spiny sharks, which had bony fin spines, growing out from the back and lower surfaces, and a skin made of minute bony scales. The details of their anatomy and their relationship to the agnathans and conodonts are not clear, but it is known that they had large eyes and jaws with teeth. Some of them were predatory carnivores,

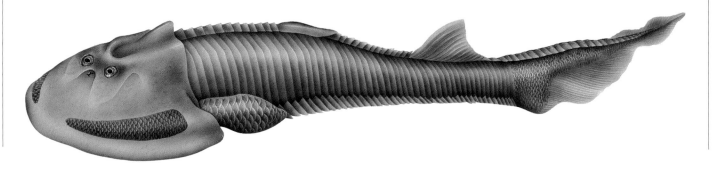

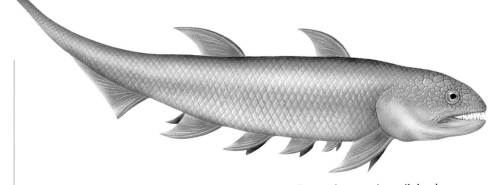

ARMED TEETH *The evolution of jaws and teeth in Silurian times allowed fish to become predators and to eat larger prey. Many had spines and tough bony scales for protection.*

and one fossil acanthodian has been found with an entire smaller fish inside. Given the fact that the underwater arms race was already well under way by late Silurian times, it is hardly surprising that fish were already arming themselves with sharp spines and tough bony scales. Exactly the same pressure of competition may also have a reason why some fish left the seas and took the perilous journey inland and into fresh waters.

The evolution of jaws was a turning point in the history of life on Earth. It promoted the struggle not only between species of animals, but also between animals and plants. Most of today's vertebrates, humans included, are dependent on their teeth. Plant-eaters, or herbivores, need teeth to remove and consume tough plant tissues, while meat-eaters, or carnivores, need teeth to catch, kill and consume their prey. The long and complicated history of vertebrate evolution can therefore be traced back to the jawed fish of the Silurian era.

GREENING THE LAND

All the organisms described so far have derived from the sea, and the Silurian beach has been our only source of evidence of life. But what of the land beyond the sand dunes? As far as the eye could see, there would have been nothing but bare grey-black rock, brown and orange mud, and yellow-white sand. The landscape would have appeared barren and devoid of life, with no trees, shrubs or grasses anywhere.

Rainwater flushes off rocky barren landscapes, such as those of the Silurian era, very quickly, removing loose rock debris, eroding

HORSESHOE HEAD *This fossil cephalaspid, found in Scotland, belonged to a genus of jawless freshwater fish, growing up to about 4 in (10 cm) long.*

the surface and preventing soil development. Most plants need soil to root in, to provide mineral salts, and to absorb the water that is so essential in preventing the tissues from be-

coming desiccated – non-woody plants may consist of up to 90 per cent water.

Despite the barren landscape and lack of soils, there is fossil evidence from the

CENTRE STAGE IN THE STORY OF EVOLUTION

In 1992, the British researcher Ivan Sansom and his colleagues, Paul Smith, Howard Armstrong and Moya Smith, helped revolutionise the early evolutionary history of the vertebrates. They demonstrated that an extinct group of tooth-like micro-fossils, called conodonts (literally, cone-teeth), are constructed of hard tissue, very similar to the cellular bone, tooth enamel and calcified cartilage found in vertebrates.

Normally, the fossil remains of conodonts consist of nothing more than minute, asymmetrical teeth – or denticles – about a millimetre or so long, which are preserved as single curved cones or as arrays of spines. These were first described as fish teeth by Christian Pander, a Russian naturalist, back in 1856. Over the next century and a quarter, conodonts became the subject of much research, as scientists sought to find a taxonomic home for them; even quite recently, they have been thought of as primitive plants.

The one crucial gap in our understanding of conodonts was exactly what kind of organism they belonged to. From the fossil evidence, the cone-shaped teeth clearly belonged to a small marine organism; and in a few rare specimens, these occurred as left and right members of a pair or pairs. Then, in 1983, a few specimens of an eel-like animal, 1½in (3.8cm) long, were found in the 'Shrimp Bed' near Edinburgh, which dates from the Lower Carboniferous era about 350 million years ago. At one end of the creature was a cluster of conodonts arranged as a symmetrical apparatus. But in these finds, the animal also had traces of a primitive backbone, or notochord, and a series of muscle blocks. Palaeontologists now had to consider the possibility that conodonts were chordates, a division of the animal kingdom that includes the vertebrates.

In 1995, the conodonts took what is probably their final step into the limelight of evolution. This breakthrough came with the discovery of exceptionally preserved specimens of 'giant' conodonts (up to 4¼in (10.8cm) long) in 440-million-year-old marine cold-water deposits in South Africa. A team of British scientists led by Dick Aldridge demonstrated that these fossils not only had the teeth forming the feeding apparatus, but also soft tissues – the oldest known vertebrate body muscles in the fossil record.

The eye muscles are particularly interesting. Eyes are not unique to vertebrates, but eyes with external strap-like muscles for rotating the eyeball are. The prominence of the eyes suggests that these creatures were predators, unlike the first agnathan fish which were toothless and probably fed on mud and algae.

Overall, these discoveries about the conodonts have transformed them from the ranks of the 'Miscellanea' of palaeontological treatises into the centre stage of early vertebrate evolution.

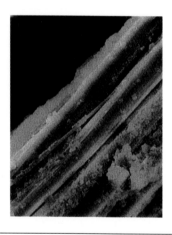

FIRST VERTEBRATES
The discovery of fossilised muscle tissue (left) provided vital clues to the conodont animal's vertebrate ancestry. The little eel-like fossil (right), found in rocks near Edinburgh, showed for the first time what the conodont animal looked like.

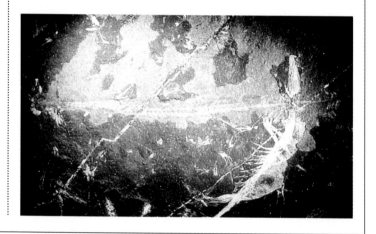

Silurian beaches that there were a few plants growing somewhere inland. Then, as now, rivers washed enormous volumes of silt and mud into the sea. Plant debris evidently got caught in their inexorable flow and was flushed downstream. When the rivers met the sea, they slowed down, and much of their floating debris was dumped along estuarine river banks. Some got carried onto the beaches of adjacent shorelines, to be stranded, buried and fossilised.

The likelihood is that these early plants, which were still dependent on water for reproduction, used the soft wet sediments of river floodplains as substitute soils. However, the proof awaits the discovery of Silurian fossil plants still rooted in the sediment.

When searching the sediments of 420 million years ago for signs of plant life, there is always the problem of knowing what to look for. Ancient plants are unlikely to resemble the highly evolved flowering plants, known as angiosperms, that are dominant today. Even apparently simple structures such as leaves are, in fact, very sophisticated. So what are the most basic and essential structures that make a successful land plant? Since a plant has to fix light energy from the Sun through a process known as photosynthesis, it is useful to grow towards the light and away from the ground surface.

Growing against gravity, however, requires a structurally reinforced, upright stem; this has the added advantage of helping the plant with the dispersal of its reproductive cells, or spores, if these are placed at the top of the stem. Even a tiny stem and a canopy of divergent branches requires a root system to anchor it, to tap water and dissolved nutrients from the ground, and to draw them into the plant. The surfaces of the plant tissues also need to be protected from drying out.

In order to survive on land, rather than in the sea which is a much more supportive medium, primitive plants had to resolve these tricky technical and mechanical problems all at once. That they did so is shown by the discovery of fossils in Silurian rocks of forked stems, only an inch or so (2.5-4 cm) in length, with spore containers at the end of the branches.

Very few Silurian plant fossils are well enough preserved to show much in the way of anatomical detail. And it has taken many years of searching and conjecture to find out exactly what kind of plants they were. The discovery of 'breathing pores', or stomata, on the stem surface has shown that they were capable of exchanging gases with the atmosphere through photosynthesis and respiration. The extra-strong cells, known as tracheids, found within their stems, indicate that the plant was capable of holding itself upright and growing away from the ground, in defiance of gravity – in a way that aquatic plants are unable to do. And the reproductive capsules have been found to be full of distinctively shaped spores.

These minute plants, called *Cooksonia*, were among the first specks of green to colour the land since the Earth had been formed, over 4000 million years before.

The mosses and liverworts that form part of a major division of plants, known as the bryophytes, are familiar to us today.

Capable of living in some of the world's most inhospitable environments, where other land plants simply cannot survive, their primitive structure suggests that they, too, may well have been among the first plants to colonise the land.

The fossil evidence, however, does not as yet support this theory. Fossil bryophytes from any time in the past are extremely rare and, until recently, the oldest found were Late Devonian in age, and no spores were preserved. Then, in 1982, bryophyte-like spores were discovered in 470-million-

year-old Ordovician rocks from Libya in North Africa. No parent plants were found until 1995, when a tiny fragment of a Devonian fossil liverwort plant was uncovered, containing spores like those from Libya. Although the fossil is some 100 million years later than the Libyan one, it does provide a link between this type of spore and the bryophytes. Patches of moss and liverworts could have grown wherever the ground was moist enough; and their decay products could have promoted the development of the first soils. From the Silurian era the greening of the landscape accelerated over the next 100 million years until plants ruled the Earth. The battle between plants and animals had just begun.

LAND PLANTS ARISE *Wet flood plains (below) were the first to be greened by the invasion of land plants. The earliest plants, like* Cooksonia *(left), grew as short, upright, leafless stems; the stem branches ended in spore-bearing capsules.*

THE INVASION OF THE LAND

One of the most momentous events in the history of evolution was the abandonment of life in water for life on land. Who first took that step and what prompted them to leave a supportive liquid for a light, dry gas?

It was during the Devonian period, between 410 and 355 million years ago, that plants gained a firm foothold on land, and that our remote ancestors – animals with backbones and four legs, known as tetrapod vertebrates – followed them out of the water. This is the sequence of events that might be expected from an understanding of how the food chain and ecological succession works. After all, plants secure energy from the Sun and thus provide food for plant-eaters, or herbivores, who can then be eaten by meat-eaters, or carnivores. But research over the last few years has shown that the story is not as simple as this and raises a number of questions.

How, for example, did these four-legged tetrapods, the 'first footers', come to develop feet and legs? And having done so, how did they make the move from sea to land? Since organisms cannot will or plan the direction of their development, the fish could not have looked longingly landwards and then developed legs to achieve their desire. Legs must have evolved while the animals were still in the water and then adapted for life on land. To answer these questions, it is necessary to understand what the land environment was like at that time, and what sort of equipment the plants and animals that left the water needed for living there. Looking at a map of the Devonian world, around 380 million years ago, can be a confusing experience. Few of the familiar parts of the global jigsaw are to be found. Neither the exact shapes of the continents, such as North America, Africa or Antarctica, are in place, nor any of the seas, such as the

PRIMEVAL SLIME *Fossils prove that aquatic algae, such as these from New Mexico, are among the oldest inhabitants of the Earth's surface.*

Mediterranean. There are oceans and continents, with mountain ranges, lakes and major rivers, but none is recognisable.

On closer inspection, however, it is possible to determine some of the pieces of this ancient jigsaw. The major continental masses of North and South America, Africa, India, Australia, Antarctica and part of Asia are all there, but they are arranged in a completely different way. This is because, over millions of years, the continents have been shuffled across the globe by forces deep within the Earth.

Such changes in the global environment have had drastic effects on life in the past. Since Devonian times, the land that forms the British Isles has moved northwards from a position 30° south of the Equator. Over a period of some 330 million years, this epic voyage has taken the region through a range of climatic zones from tropical to polar. All the generations of different creatures, including humans, who have occupied the region throughout this period have had to adapt to the changing environment – or become extinct.

Two major continental groupings dominated the Late Silurian and Early Devonian world map. The southern continents of

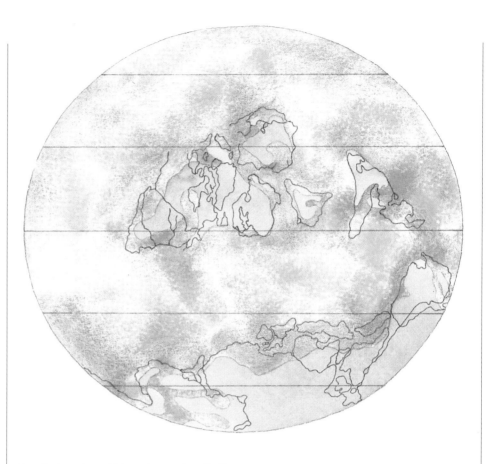

South America, Africa, Antarctica, India and Australia, were all clustered together to form the supercontinent known as Gondwanaland. The other supercontinent is known as Laurasia and comprised North America and northern Europe, which had gradually moved together during the Lower Palaeozoic era. In the process, the continents had closed the Iapetus Ocean, which had separated them for the previous 200 million years, much as the Atlantic does today. The effect of two major continental masses being driven together was earth-shattering. As the continents collided, the relatively soft sedimentary rocks that fringed the land like aprons, spreading out into the shallow seas, bent and buckled into great folds of mountains, up to a mile or so high and hundreds of miles long.

Despite more than 380 million years of weathering and erosion, it is still possible on today's maps to identify this ancient line of impact in the range of mountains thrown up by the collision of North America, or Laurentia, and northern Europe, or Baltica. From northernmost Norway, a mountain

PUZZLING PIECES *The plate tectonic movement of the continents has resulted in a very different global jigsaw from 380 million years ago.*

chain runs south-west through Norway, across to Scotland and northern England, over to northern Ireland and south-west from Connemara out into the Atlantic. Some 3000 miles (4800 km) across the ocean, the broken end of the mountain chain can be picked up once again in Newfoundland, continuing its south-west course through New England into the Appalachians.

A NEW LANDSCAPE FOR LIFE

The emergence of this landscape from the waters of the Silurian seas coincided with some massive evolutionary developments. As yet, there is little evidence of any substantial colonisation of the land before the Devonian era. Fossil discoveries have shown that the first advanced land plants had evolved over 400 million years ago. But it is *continued on page 34*

The Late Devonian Landscape

Reconstructions of the ancient world, like this Late Devonian landscape, are the culmination of over a century of investigation by palaeontologists around the world. Once fossils have been discovered, the scientists examine their structure in detail, analyse their function, and try to relate them to the environment in which they are thought to have lived.

New discoveries of creatures such as *Acanthostega*, and the reinterpretation of *Ichthyostega*, have shown that the early tetrapods were mainly aquatic, and much more at home in water than previously thought. The development of their limbs may therefore have more to do with survival in freshwater rivers than with occupying the land. Nevertheless, the possession of limbs may have prepared the tetrapods for life on land.

By the Late Devonian era, vascular plants were taking root beyond the watery confines of marshes and river banks. They included the first trees with woody tissues strong enough to support leaf canopies many feet above the ground. Fossil evidence shows that terrestrial food chains were well established by this time. Debris and litter from plants, including pioneer horsetails, fern-like plants and lycopod club mosses (some of which grew to heights of 60 ft/ 18 m), were broken down by fungi and bacteria, and eaten by arthropods, such as the early millipedes. These, in turn, were preyed upon by the first large, land-going scorpions.

1. *Archaeopteris*
2. Scorpion
3. *Sciadophyton*
4. Horsetails
5. *Aglaophyton*
6. *Acanthostega*
7. *Ichthyostega*
8. *Cyclostigma* (lycopod club moss)
9. Cladoxylon (early fern)
10. Asteroxylon
11. Millipede

THE FIRST PLANT-EATERS

The first evidence of animals eating plants comes from Late Silurian and Early Devonian fossils, around 400 million years old. In 1995, a research team from the University of Wales found minute fossilised pellets in some ancient sedimentary rocks from the Welsh Borders that had been dissolved in acid in a laboratory in Cardiff.

When magnified 500 times with an electron microscope, the nature of the pellets was not immediately clear to the scientists. The tiny pellets consist of densely packed plant spores, and at first it was thought that they were the isolated reproductive organs, or sporangia, of primitive land plants such as

Cooksonia. But then it was realised that each pellet contains up to nine different kinds of spore, each one representing a different plant.

Plant scientist Dianne Edwards showed the pellets to a colleague, Paul Selden, an expert on fossil arthropods at Manchester University. He realised that they resembled the fossilised faeces, or coprolites, of living myriapod arthropods (invertebrate organisms, such as millipedes, consisting of many segments each bearing pairs of legs). Apparently, some myriapod-like arthropod had found a way to survive on land by browsing on the sporangia of early land plants.

VEGETARIAN FIRST *The earliest evidence of animals eating land plants comes from a minute fossil dropping (left). At high magnification (below), the pellet – or coprolite – can be seen to contain hundreds of plant spores.*

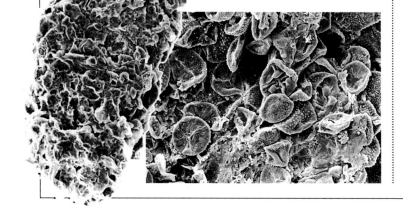

Scientists soon realised that they had stumbled upon the first evidence for plant-eating, or herbivorism, in the fossil record.

However, the question of whether the spores were eaten live on the plant or were picked up from the decaying plant litter on the ground still remained. By increasing the microscope magnification to over 1000, it became clear that the spores were very well preserved, whereas plant litter on the ground decays and is broken down by fungi and bacteria. This suggested that the spores were eaten fresh, while still on the living plant. Furthermore, the herbivore must have been feeding on the tissues of the sporangia, which cover the spores, because the spores have passed through the animal's gut without any sign of digestion; in other words, the spores were just 'roughage'.

PROBABLE PIONEERS *Eaters of plant debris, like the millipede (above), would have been well adapted for life on land, but they were not alone. Stem sections of fossil plants found in Rhynie chert (below) show the scars of wounds, originally made by unidentified sap-sucking arthropods.*

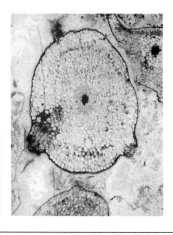

likely that these minute green plants were restricted to damp environments, creating little more than a fuzzy green 'halo' along muddy river banks. The arid interiors of the Laurasian supercontinent, draped across the Equator, were still harsh and hostile environments for life. It was not until the Devonian era that all this was to change.

To survive on land, any organism emerging from the sea had to be able to withstand heat and desiccation, damaging ultraviolet light, and an atmosphere with

some ten times as much carbon dioxide as today. If it wanted to move, it had to defy gravity and to overcome surface friction, in a thin unsupportive gas – the atmosphere. It would also have needed sensory equipment, like sight, smell and hearing, that would work in the dry. And, most important of all, it would have needed food.

Looked at in this way, it seems remarkable that any creatures bothered to leave the sea. But the increasing intensity of predator-prey relations in shallow seas had,

by Silurian times, turned the waters into a battleground. By comparison, the wide open spaces and unoccupied niches of the land were up for annexation by any pioneer colonists who were tough enough and properly equipped to cope with the hardships the land presented.

Ultimately, all life depends upon plants. This is because the food chain begins with a basic source of energy – light energy derived from the Sun. The organisms best equipped to convert light into chemical energy are the plants, through processes known as photosynthesis and respiration. You would therefore expect plants to have preceded animals in the invasion of the land. And, indeed, this is what the fossil record suggests: with the appearance of *Cooksonia* in Silurian times, more than 400 million years ago.

FINDING THE FIRST PLANTS

Tiny fragments of plant-like fossils have been found in the Welsh Borders for over 150 years; and it was here, in 1990, that researchers unearthed the oldest fossils of a true land plant ever found. They were looking for the presence of special 'breathing' pore cells, or stomata, in the plant skin, or cuticle, and strengthened supporting cells, or tracheids, in the stems. These were features that would have enabled the plant to

*RHYNIE PLANTS   These three Early Devonian bog plants – *Rhynia *(left),* Asteroxylon *(centre) and* Aglaophyton *(right) – released their spores into the wind from structures known as 'terminal sporangia' at the tips of their stalks.*

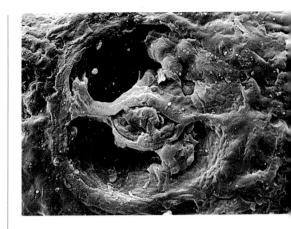

BREATHTAKING   A stomata cell, on the stem of a fossil Cooksonia, *allowed the early land plant to 'breathe' the Silurian air.*

conduct water and nutrients from the ground through the stem tissue, thereby enabling it to grow upright. The *Cooksonia* the researchers found, in a small quarry near the village of Stoke Erith in Herefordshire, had beautifully preserved stomata on the stem surface and tracheids inside, and dated from around 410 million years ago. Furthermore, the spore-bearing capsules at the end of the forked stem were intact and full of spores, which linked them with a particular *Cooksonia* species, *Cooksonia pertoni*.

More detailed information about early land plants comes from the fields around Rhynie, a village in north-east Scotland. In 1913, a source of fossils was discovered by accident, when blocks of a hard flinty rock, called chert, were found in a stone wall surrounding a farmer's field. Closer inspection showed that these contained plant fossils, including a *Cooksonia* species. Because chert is like a glass and can be cut with a diamond rock saw into very thin translucent slices,

and because the rock preserved the plants before the normal processes of fossilisation could flatten them or destroy their internal structures, these 390-million-year-old plants can be viewed at a level of anatomical detail that is normally only found in botanical preparations of living plants.

To find the source of the plant-rich rock, geologists this century have dug trenches in the fields around Rhynie. These show that the chert was interlayered with sand and mud sediments of Early Devonian times. Twenty-two different primitive plants and several terrestrial arthropods have been identified, all of which are unique to

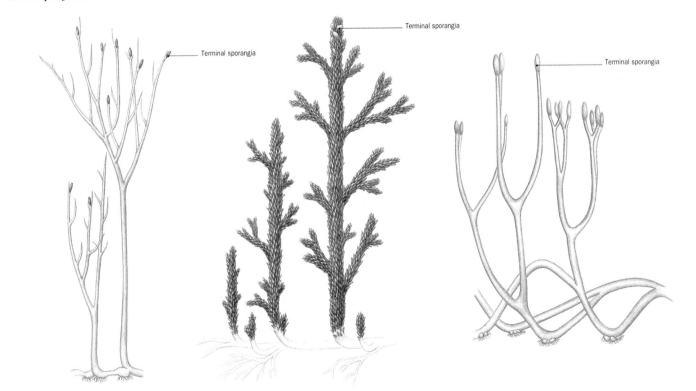

Terminal sporangia

Terminal sporangia

Terminal sporangia

Rhynie. Taken together, and including the algae and fungi, they represent the oldest plant ecosystem – a complete ecological community – about which much is known. Some of the plants even have wounds, typical of those produced by sap-sucking insects today. Perhaps the most important of the Rhynie plants is the world's first known club moss, or lycopod, *Asteroxylon*.

PLANT PROLIFERATION

Within less than 50 million years, all the major groups of vascular plants (those with vessels for carrying sap) had evolved from these simple beginnings – except for the flowering plants, or angiosperms. There must have been a very rapid greening of the land, with the development of horsetails, fern-like plants (growing to 3 ft (1 m) or so in height by Mid Devonian times), plants with distinct foliage, seeds for propagation, and even the first tree-like plants.

All these developments required further structural innovations. Stem tissues had to become a lot more complex to maintain the vertical growth of plants that were reaching heights of 70 ft (21 m) or more by the end of the Devonian era, when the first forests developed. And special tissues had to be developed to enable the transfer of energy and water through the plant, from the highest branches to the lowest roots.

The most significant development of all, perhaps, was the evolution of seed plants, or gymnosperms. All the other primitive plants relied upon fertilisation by the male reproductive cells, or gametes, moving through water towards the female reproductive organ. In the gymnosperms, which developed at the end of the Devonian era, the male reproductive cells – in the form of pollen – were carried in the wind and fertilised the female cells. Seeds developed, which were then in turn distributed by the wind. This innovation released them from the constraints of having to live in moist

FOSSIL TRACERY *A Devonian mudstone shows that fern-like plants, such as* Archaeopteris, *soon evolved the ability to grow stems and leaves.*

plant with the help of bacteria in its gut. These gut bacteria would first have been acquired as animals fed on plant debris that was itself in the process of being broken down by fungi and bacteria.

THE ANIMALS MOVE UP RIVER

Like any sensible explorers of inhospitable terrain, animals first penetrated the interior by river. Even to do this, they had to cope with the chemistry of fresh water. Seawater has a similar saltiness, or salinity, to cell fluids within an animal's body, with the result that there is a natural balance. But there would have been a marked salinity imbalance when animals entered fresh water for the first time; and this would have caused stress, shock and damage as the different fluids leaked through the animals' semipermeable skin.

What was needed was a variety of water-proofing devices and a way of getting rid of excess water – such as the kidney in the case of the first vertebrates.

Early animal explorers needed to be mobile and to have adapted to such salinity changes. The most likely candidates were those which had lived in and around the deltas and estuaries of major rivers, where saline levels were slightly lower. Arthropods, such as the millipedes and eurypterids with their tough exoskeletons, were amongst the best-equipped to pioneer the perilous inland routes, followed by the fish, and subsequently the bivalved molluscs. Many of these animals must have been feeding on freshwater plants, algae, other microscopic plant organisms, and their decay products. Next came the carnivores – in the form of

environments for successful sexual reproduction: plants were able to invade really dry land for the first time and move towards the hills. This technical feat kept the plants one step ahead of the invading animals. Such innovations set the stage for one of the most remarkable and successful takeover bids made by any organism: the domination of the land by the luxuriant Carboniferous floras.

So far, the invasion of the land by plants has dominated the story of the Devonian era – an argument justified by the fact that plants lie at the base of the terrestrial food chain. Carnivores depend on herbivores, who depend on plants. But plants are not a defenceless 'soft option', just waiting to be eaten by herbivorous animals. Land plants are, for the most part, fairly indigestible; and to obtain energy from the plant, the animal has to be able to break down chemically the tough cellulose cell walls of the

THE FIRST FOOTPRINTS

The world's oldest terrestrial footprints were found in 1993 by British geologists, within the 450-million-year-old Ordovician rocks of England's Lake District. They consist of two parallel lines, like tank tracks but less than half an inch (1.25 cm) apart. Each line consists of tiny pairs of triangular indentations, made by a number of pointed feet.

In the absence of any fossil remains of their bodies, the creatures responsible for these footprints could only be identified by analysis of the trackways and by comparison with similar tracks made by living organisms. Palaeontologists from Bristol University showed that the tracks were made by multi-legged creatures, whose paired limbs were strong enough to lift their bodies clear of the soft sediment. The number of limbs indicated that they must have been

segmented, rather than soft-bodied, creatures, probably myriapod arthropods which are rather like modern centipedes. These ancient creatures inhabited a constantly changing environment of shallow lakes and rivers in the days when the Lake District was part of an active island arc of volcanoes.

Spectacular trackways are also preserved in the Silurian sandstones of Western Australia, which are

younger than the Lake District rocks by some 30 million years. Scientists showed that the largest of these trackways, which are 8 in (20 cm) wide, were made by eurypterids up to 6 ft (1.8 m) long. Smaller tracks were made by another extinct arthropod group of cockroach-like creatures, called euthycarcinoids, which were 2 in (5 cm) long, had 10 or 11 pairs of legs, and were related to the centipedes.

All these fossil trackways indicate that the arthropods were able to move into fresh water and to survive exposure to the atmosphere much earlier than had previously been thought. It was this ability that clearly provided the critical stepping stone between the sea and complete terrestrialisation.

LIFELINES *Some 450 million years old, these fossilised parallel trackways are the first evidence of animal life on land.*

jawed fish, which had inhabited the Siluri-an seas but were introduced into the fresh waters for the first time in the Devonian era.

One problem of living in river systems is that they are so ephemeral. Water levels, which depend to a great extent on seasonal rainfall, can fluctuate enormously from flooding to drying up completely. Rivers also change course as they develop. Animal

LAND INVASION *Scientists are still puzzling over how and why vertebrates left the water for the much harsher environment of life on land.*

survival in this environment depended on mobility and, often, on an ability to cope with low levels of oxygen in the water – any creature that could get oxygen directly from the atmosphere would have been at a distinct advantage in the unpredictable habitats of the Devonian period.

In most respects, the common amphib-ians of today, such as salamanders, frogs and toads, are equally at home on land and in the water. However, amphibians need to return to water to breed. Many living rep-tiles, on the other hand, are more fully adapted to life on the land, for although turtles and alligators live happily in water,

they are dependent on land for breeding.

It would be expected, therefore, that in the evolution of the vertebrates, the am-phibians preceded the reptiles in the fossil record. Both these major groups are distin-guished from the fish in having two pairs of limbs for movement; they have four feet and are known as tetrapods.

It might seem surprising that the oldest known tetrapod fossils were found in the icy wilderness of Greenland. But in Devon-ian times, Greenland was part of the great new Laurasian continent and lay much far-ther to the south than it does today. In what must seem like the most extraordinarily

The first steps towards land vertebrates started way back in the seas of Cambrian times. From small eel-like lancelets and conodonts, with the beginnings of a flexible backbone, the vertebrates evolved through a wide range of bizarre jawless and toothless fish types during Ordovician and Silurian times (510-410 million years ago).

Silurian acanthodians were the first jawed fish with teeth and sets of paired fins stiffened by sharp spines.

In Early Devonian times, the fish first invaded fresh waters and the lobe fins, such as *Eusthenopteron,* had two sets of strong muscular 'shoulder' and 'hip' fins. These paired fins provided the basis for all subsequent tetrapod limbs, such as the Upper Devonian *Acanthostega* and *Ichthyostega,* the first vertebrates capable of moving out of water and supporting themselves on dry land.

Promissum
(conodont)

Errivaspis
(pteraspid)

Hemicyclaspis
(cephalaspid)

Jamoytius
(anaspid)

Mesacanthus
(acanthodian)

Astraspis
(astraspid)

serendipitous find, fossil bones were recovered here in 1897 by Danish explorers searching for a missing explorer, Salomon Andrée, who had tried to reach the North Pole by hydrogen balloon.

THE FIRST FOUR-FOOTERS

It took another 30 years before the significance of this find was realised. From 1931, the Swedish palaeontologist Gunnar Säve-Söderberg returned each summer to the 360-million-year-old sandstones, high in the snowy wastes of eastern Greenland. Several skulls were found, but it was not until 1948 that Säve-Söderberg's assistant, Erik Jarvik, found most of the rest of a skeleton: the oldest known backboned animal with four well-developed limbs. Named *Ichthyostega*, the creature was 3 ft (1 m) long and resembled a salamander; it had lived and died in one of the ancient river systems that flowed across the great Devonian landmass. Jarvik continued to work on some details of *Ichthyostega*'s anatomy until the 1990s. Meanwhile, this ancient creature's story had been partly overtaken by finds of even more primitive tetrapods in East Greenland.

During the short Arctic summer of 1987, Cambridge scientist Jenny Clack visited Greenland and discovered the fossil remains of another 3 ft (1 m) long, salamander-like amphibian, called *Acanthostega*, which was even older than *Ichthyostega*.

Acanthostega, like *Ichthyostega*, retains a number of fish-like features, in particular its long, flattened tail and internal gills. Even the details of its limbs suggest that *Acanthostega* was better adapted for swimming – grubbing about in soft sediment for

From animals such as these, the early tetrapods rapidly evolved with numerous crocodile-like amphibians and, during the Carboniferous era, the first reptile tetrapods such as *Hylonomus*.

Ichthyostega
(early tetrapod)

Hylonomus
(reptile)

Eusthenopteron
(lobe-finned fish)

Acanthostega
(early tetrapod)

food or pushing itself along in very shallow water – than for walking on dry land. Nevertheless, the possession of four limbs is also just right for scrabbling across wet mud from one pool to another in a drying watercourse; and limbs like this could also help an animal chase small prey out of the water.

One of the most surprising features of *Acanthostega* was revealed when Jenny Clack and her assistants removed the hardened sediment from around the limb bones. They were amazed to find that the animal had eight fingers or seven toes on each limb, rather than the five fingers or toes that biologists had previously assumed were characteristic of the earliest tetrapods.

By the Late Devonian era, when these first fossil tetrapods evolved, fish swarmed in both the seas and fresh water. Jawless and jawed fish had evolved into many different groups, exploiting a variety of aquatic niches. Only the jawed fish had the potential to become predators, while the jawless fish had to rely on armour and size for defence.

Of all the numerous groups of bony jawed fish, only a few had the potential to evolve into tetrapods. Tetrapods need internal supports, or bones, and muscles for the limbs to be able to raise the body off the ground. These supports have to be firmly attached to the main structural bridge, the backbone, which supports the body. The body of a land-going tetrapod vertebrate can therefore be viewed as a kind of suspension bridge, in which the main mass of the body hangs from a horizontal flexible beam, the backbone, which is in turn held up by two pairs of legs. In tetrapods, the backbone is connected to the appendages

PRIMITIVE TETRAPOD *Fossil remains of* Acanthostega *(below), found in Greenland, have enabled scientists to reconstruct the creature (right). Each of its limbs was equipped with seven or eight digits, with which it hauled itself up muddy river banks.*

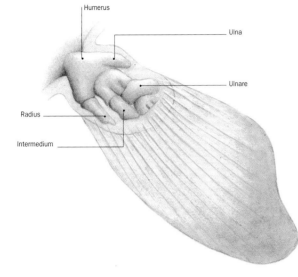

Humerus
Ulna
Ulnare
Radius
Intermedium

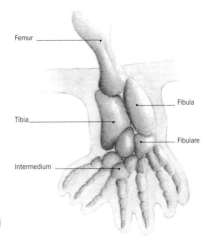

Femur
Fibula
Tibia
Fibulare
Intermedium

FROM FIN TO FOOT
By comparing fossil bone patterns, palaeontologists have worked out how the first tetrapod limbs, such as that of Ichthyostega *(left), evolved from the fins of lobe-finned Devonian fishes (far left).*

landmasses, moving upstream and feeding off the aquatic plants and invertebrate creatures that had pioneered these freshwater routes. Life became increasingly competitive – a problem compounded by the unpredictability of an environment, in which rivers kept changing their levels and their courses. To survive, it became an advantage to possess muscular limb-like paddles. These enabled some creatures to step out of the water and onto dry land – if only, at first, to escape predators, to pursue prey or to move from one watercourse to another. The land-going tetrapods had arrived and the story of prehistoric life took a great step forward.

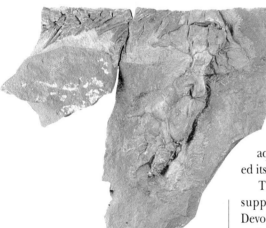

HIND LIMB *By studying the fossilised limbs of the oldest tetrapods, such as* Ichthyostega *(above), scientists concluded that the primitive creature had more than five toes.*

tetrapods and the bony fish is made by comparing the fossil fin bones of a particular form of osteolepiform, called *Eusthenopteron*, with the limb bones of the tetrapods. But there remains a substantial gap between the two groups: *Eusthenopteron* was still an entirely aquatic fish and could not have supported its body out of water with its fins.

The link – however imperfect – does support the argument that, during the Devonian era, new groups of predatory fish evolved to dominate the seas. They invaded the rivers and lakes of the newly emerging

FRESHWATER FISH *With its paired fins,* Osteolepis – *preserved here in sandstone – was one of many kinds of Devonian freshwater fish.*

by pectoral and pelvic girdles – shoulders and hips – which enable the bridge to walk.

All these structures represent a considerable advance on the basic fish skeleton. However, a few groups of Devonian jawed fish possessed features that show developments in this direction. Pairs of long fins, with strong internal supporting bones and girdles for attachment to the backbone, are found in lobe-finned fish, such as the lungfish, coelacanths and an extinct group called the osteolepiforms. The closest link that can be established at the moment between the

THE ANCIENT FORESTS

The first forests to green our planet helped to oxygenate the global atmosphere. Their fossil remains also formed the coal deposits which, when burned in vast quantities each year, are changing the atmosphere once again.

The first large forests on Earth grew during Carboniferous times, from 355 to 290 million years ago. Over the last 200 years, the fossil fuels derived from them in the form of coal and gas have fuelled the Industrial Revolution and the subsequent economic development of the modern world. As coal mines were dug, the fossil record was uncovered to reveal an enormous amount about the animals, plants and their environments during the Carboniferous era – a geological period that literally means 'coal-bearing'.

The coal swamps and forests of the Carboniferous era provided new opportunities, new habitats and new sources of food, and they soon teemed with animal life, from insects to amphibians and reptiles. But as well as the flourishing tropical forests of the Carboniferous era, whose plants eventually formed coal, there were other environments too.

The Early Carboniferous era, or Mississippian as it is known in North America, was a period of rising sea levels. Many of the low-lying coastal regions of the old Devonian landscape were flooded and turned into shallow water, or continental shelf seas. In equatorial regions, covering what are now America and much of Western Europe, the shallow tropical seas were ideal for the development of coral reefs – on a vastly greater scale than even the Great Barrier Reef of today. These ancient seas were crowded with all kinds of newly evolved creatures, from shellfish and corals to a variety of bony and cartilaginous fish, including an array of primitive sharks.

Recent discoveries in the east of Scotland have provided a remarkable insight into the kinds of plants and animals that occupied the Early Carboniferous landscape. At the time, Scotland was an upland area, part of a mountain belt that ran from Scandinavia to North America. The site on which Edinburgh now stands was originally located well inland, on the forested

FOSSIL STING *There is no doubting the purpose of the curved spine on the tail of Pulmonoscorpius. By Carboniferous times, scorpions had adapted to life on land.*

LIVING COAL *Modern tree ferns are a living reminder of what Carboniferous 'Coal Measure' – or coal swamp – forests were like.*

flanks of that mountain chain. Edinburgh Castle is strategically placed on top of Castle Rock, the remnant of a Carboniferous volcano.

WINDOW ON A WORLD

Some 338 million years ago, this volcano dominated the area, intermittently belching clouds of ash and spewing lava over the landscape. Groundwater heated by the hot rocks within the Earth broke through to the surface, producing hot mineral springs and lakes. Ash was washed off the flanks of the surrounding hills and volcanoes, and carried into the lakes by streams. Although the water in these lakes was toxic enough to kill many of the creatures that entered it, it did provide an excellent medium for preserving their remains as fossils.

The sediments deposited in one of these small lakes, at the base of the volcano, are still preserved today – as limestones in the old quarry of East

Kirkton, in the Edinburgh suburb of Bathgate. In 1984, Stan Wood, a professional fossil collector, spotted the 338-million-year-old limestone rock in a stone wall while he was refereeing a local football match. Over the next few years, he systematically took some of the wall apart, stone by stone. His hunch paid off in 1988, when he discovered an 8 in (20 cm) long lizard-like fossil tetrapod. *Westlothiana lizziae* – or 'Lizzie the lizard', as the fossil was nicknamed – became internationally famous and the subject of intense speculation about reptile evolution. A second specimen of the creature

EARLY REPTILE? *Heralded by some as the oldest fossil reptile, detailed research now suggests that* Westlothiana lizziae *is a more primitive tetrapod.*

was also found. Superficially, it resembles a small lizard (although its pedigree is not yet proven), with a long tail, small teeth and limbs sticking out to the side; it may have lived in the leaf-litter on the forest floor, hunting small soft-bodied invertebrates.

East Kirkton has also yielded a bonanza of fossil arthropods, including eurypterids, scorpions, millipedes and even the oldest known harvestman spider, all well preserved in the fine-grained limestones. The scorpions are perhaps the most interesting specimens, as they are the earliest fully terrestrial ones, growing up to 11 in (28 cm) long and capable of breathing air directly. Until the finds, palaeontologists had assumed that Carboniferous scorpions were aquatic, but *Pulmonoscorpius* from East Kirkton has structures called 'book lungs' for breathing air: delicate, gill-like respiratory surfaces protected within pockets under the exoskeleton.

continued on page 46

THE FIRST FORESTS

The equatorial rain forests of Carboniferous times were the first extensive forests on Earth. By photosynthesising light energy from the Sun, the plants built their carbon-based tissues, effectively locking up vast tonnages of carbon from the surrounding environment. When the plants died, the decaying litter built up in alternating layers of peat, sand and mud. Over time, the accumulated sediment was compressed and transformed into coal seams, which have been widely exploited as fuel since the 18th and 19th centuries.

The development of the forests also altered the composition of the atmosphere itself by considerably increasing the proportion of oxygen in it relative to carbon dioxide. Burning coal in modern industrial times has unlocked the Carboniferous fossil fuel store and returned the carbon dioxide to the atmosphere. As a result, the composition of the atmosphere is once again changing and thereby helping to fuel global warming.

Carboniferous plants were largely confined to wet lowland areas, such as inland and seaside swamps. Giant club mosses, such as *Sigillaria* and *Lepidodendron* (diamond patterned trunks), horsetails (*Calamites*, sectioned trunks), ferns (*Psaronius*) and conifer-like *Cordaites* formed the high tree canopy in these forests. Vine-like seed ferns hung from the trees, while others formed ground cover along with the true ferns.

Giant millipedes (*Arthropleura*, lower centre), centipedes (on fallen horsetail) and cockroaches (*Archimylacris*, climbing a club moss) scavenged decaying plants, while spiders (*Graeophonus*, bottom right), giant dragonflies (*Megatypus*) and the first true reptiles (*Hylonomus*, on fallen horsetail trunk) hunted through the shady undergrowth. The water was home to a variety of invertebrates, from shellfish to predatory scorpions (*Paraisobuthus*, bottom left), vertebrates such as the amphibian *Dendrepeton* (bottom right) and many kinds of fish.

ANCIENT MOSS *A twig of club moss, with its distinctive numerous stem leaves, is preserved as a fossil from a Carboniferous wetland.*

The scorpions also had legs for walking and acute vision for daylight hunting.

Many amphibians have been found, too, including the oldest known member of a group called the temnospondyls, which survived for 150 million years into the Early Cretaceous period, branching into over 30 families and 170 known genera. Most of the East Kirkton temnospondyl fossils are of an animal called *Balanerpeton woodi*, which grew to about 20 in (51 cm) and probably lived most of its life on land, only returning to the water to breed. It has a long, slender earbone for transmitting high-frequency vibrations from the eardrum to the inner ear – a sure sign of an animal which needs to detect airborne (rather than water-borne) sound. Also, as in most terrestrial

animals, its ankle and wrist bones are fused for extra strength. All in all, *Balanerpeton* was very much like the modern amphibian salamander, which lives most of its adult life on dry land.

Since the original discovery, East Kirkton quarry has been searched by teams of international experts. They have investigated the fauna and flora, including ferns and club mosses that dominated the environment. Their analysis has made East Kirkton one of the most important palaeontological sites in the world.

INSECT LIFE

Palaeontologists have realised in recent years that such sites may provide 'windows' on past life that are not normally available: for example, on the early history of the insects. Insects are by far the most successful macroorganisms – many-celled organisms visible to the naked eye – of all time. Their total global population is estimated at around 1×10^{18} (a million million million), grouped into about a million known species out of a possible total of between 3 million and 30 million species.

The first true insects are preserved in the fossil record of Carboniferous rocks, although it is likely that they originated well before this and that earlier fossils have not been found yet. One of the problems in tracing the early history of the insects lies in

the fact that few of them have been preserved in the fossil record. Only certain groups, such as the beetles which have relatively tough exoskeletons, stand much chance of preservation. Secondly, most insects are small and represent relatively high-protein food – alive or dead – for a vast range of predatory or scavenging animals and, even, plants. Thirdly, for an insect to be preserved as a fossil, it needs to have been carried, usually by water, and then deposited as part of a load of sediment,

WETLAND INHABITANTS *An arthropod fossil, preserved in siltstone, exhibits the many segments and jointed legs of a millipede (left). Another creature of the Carboniferous swamps was the now-extinct amphibian* Balanerpeton woodi *(right), which lived mostly on land.*

typically in a drainage basin. Insect skeletons, however, are fragile and easily destroyed by fast-flowing rivers, with the result that it is only in slow-moving, muddy environments that there is much chance of their preservation. It is to this type of sedimentary rock that palaeoentomologists, scientists who study fossil insects, turn.

Systematic collecting from an old colliery tip at Lower Writhlington, in southwest England, has provided more than 300 new specimens of fossil insects. The fossils were found on the surface of thin layers of shale that had once been fine-grained mud, deposited by flood waters on top of the peat that eventually turned into coal.

Most of the insect fossils are the fragmented remains of various exoskeleton parts. Of these, the most useful for identification are the wings, with their distinctive patterns of veins. One of the reasons that these fossils have so often been missed in the past is that the wings look remarkably like leaves. This resemblance, especially to fern leaflets, is probably no accident, since

---

THE OLDEST DRAGONFLIES

In 1996, two German palaeoentomologists discovered a fossil of the oldest known dragonfly in Lower Carboniferous rocks in Germany. Called *Delitzschala bitterfeldensis*, it was quite small, with a wingspan of less than an inch (2.5 cm), and although some 320 million years old, the irregular colour markings on its wings were still well preserved. These would have served as camouflage for its outstretched wings and suggest that, right from an early stage in their evolution, the dragonflies had enemies – not aerial ones but land-based ones.

---

it would have given the insects protective camouflage from the increasing numbers of hungry amphibians that inhabited the Carboniferous swamps. These wing vein patterns can be used like 'thumbprints' to identify individual species. The vast majority of Carboniferous insects turn out to be cockroaches, one of the most ancient and successful groups of insects, and they have

SUCCESSFUL SURVIVORS One of the most successful insects, the cockroach, originated in the Carboniferous and has since evolved into over 2000 species.

survived virtually unchanged in form for over 300 million years. However, perhaps the most interesting and spectacular of these early insect groups are the first dragonflies. Some had wingspans of more than 2 ft (61 cm), which make them by far the largest winged insects to have conquered the air.

INTO THE AIR

One of the greatest evolutionary innovations in the story of life on Earth has been self-propelled flight – 'invented' by the insects more than 325 million years before Orville and Wilbur Wright took to the air. American biologists James Marden and Melissa Kramer developed a theory to explain just how those insect pioneer aviators took off so long ago. They proposed that wing-powered water skimming, as practised today by the stoneflies of our ponds and rivers, holds the clue as to how it happened.

Marden and Kramer suggested that the aquatic ancestors of the modern stoneflies developed small, rudimentary wings for skimming. They tested the idea by trimming the wings of stoneflies to different lengths and found that, even at 20 per cent of the original length, stoneflies can still skim – although the bigger the wings, the better they perform. The biologists argued, therefore, that the wings of the early insects gradually became bigger and better – until they evolved into structures preadapted for flight. The proof of their theory, however, still depends on finding fossil evidence of primitive stoneflies with skimming 'wings' from Carboniferous times.

Insects were able to dominate the air of the Carboniferous era because there was no

competition from flying vertebrates. Fossils from this period include the remarkable *Megatypus*, the giant predatory dragonfly, with a wingspan of some 28 in (71 cm), which was first described by the French naturalist Charles Brongniart in 1885. The dragonflies' dominance was short-lived because they lacked a technical device crucial for aerial warfare. Dragonflies could not, and still cannot, fold their wings back when at rest. Today over 99 per cent of all living winged insects can fold their wings, which both protects the wings and greatly improves their manoeuvrability on the ground, allowing them to hide from their enemies. Once most of the flying insects had overcome this hurdle, there was no holding them – at least, not until the flying reptiles began to provide some competition for airspace. By then, most insects could land, fold their wings, and hide from their pursuers, whether flying reptiles or birds.

Birds and insects evolved at the same time as plants, although at this relatively early stage in the history of life, more is known about the plants. By Upper Carboniferous times – or the Pennsylvanian era, as it is known in North America – plants had really rooted themselves in the lowland landscapes of the world.

THE PLANTS OF THE CARBONIFEROUS

Many of the dominant groups, the club mosses (lycopsids), the horsetails (equisitopsids), the ferns and the seedferns (pteridosperms) had evolved in the Devonian era. Many of them still have living representatives, some of which – like the ferns – are very successful. However, the form and size of their Carboniferous ancestors did not resemble their living descendants, and as a result the woods and forests of 300 million years ago would have looked totally different from those of today.

The club mosses evolved in the Devonian era, soon after plants had colonised the land, flourishing in low-lying, swampy ground. By the Upper Carboniferous era, club mosses had evolved into giant forms

MODERN MINIATURES *Living horsetails (*Equisetum*) are minute, compared with their tree-sized Carboniferous equisitopsid ancestors.*

such as *Sigillaria*, which grew to 100 ft (30 m) and had few branches, and *Lepidodendron*, which reached 165 ft (50 m) and branched to form a dense canopy. These giant club mosses died out in Permian times (some 260 million years ago), but some smaller forms survived, eventually giving rise to the tiny quillworts that live today in a range of freshwater and terrestrial environments.

Like the club mosses, the horsetails first evolved in Devonian times and reached their fullest development, both in size and in numbers of species, during the Carboniferous era. The largest of the horsetails were the *Calamites*, some of which grew as high as 65 ft (20 m). Spreading out from long underground stems, which penetrated the soft muds and sands of river deltas, the tall, upright stems of the plants branched out to form dense thickets, like bamboo groves. *Calamites* and *Sphenophyllum*, another

horsetail-like plant, both died out at the end of the Carboniferous era, but other horsetails survived to form a group of 20 or so species of *Equisetum* in Triassic times (about 230 million years ago). These plants are still found around the world, although none of them reaches the tree size of their Carboniferous ancestors.

The true ferns also originated in Devonian times and multiplied into an array of Carboniferous forms, from which some of today's tropical ferns are descended. Many of these Carboniferous ferns grew into tree-size plants 60 ft (18 m) or more high, with individual leaf fronds over 10 ft (3 m) long: *Psaronius*, which is related to the living tropical fern *Marattias*, is one such example. Superficially, these giant ferns looked like modern tree ferns, such as *Dickinsonia*, but their reproductive characteristics were

FIRST FLIERS *One of the first creatures to take to the air was a dragonfly, called Namurotypus. This specimen was preserved as a fossil in the Carboniferous lake mud.*

N 1041
20.12.1985

rather different, with large spore cases fused in clusters on the underside of the fronds. Unlike the horsetails and club mosses, the true ferns evolved beyond the Carboniferous era. Today, there are over 12 000 living species of ferns to be found in virtually every plant habitat around the world.

At first, the leaves of the seed ferns look deceptively like those of the true ferns; indeed in the last century, many of the fossil leaf fragments found in the Carboniferous coal seams were mistaken for true ferns. What distinguishes them, however, is that the fronds of most seed ferns have more complex branching patterns, and larger and tougher leaflets. The seed ferns of the Carboniferous era included ground creepers,

climbers or lianas, shrubs and tree-sized plants up to 33 ft (10 m) high. Their leaves were up to 23 ft (7 m) long and formed enormous canopies. One of the groups of these ancient seed ferns, the trigonocarpaleans, may have been the ancestors of the cycads, which flourished in the Mesozoic era, during the time of the dinosaurs,

CHANGING PLANTS *Tree-sized ferns, like the living* Cyathea *found in a New Zealand rain forest (below), were a feature of Carboniferous forests. So were seed ferns, from which today's cycads (right) may be descended.*

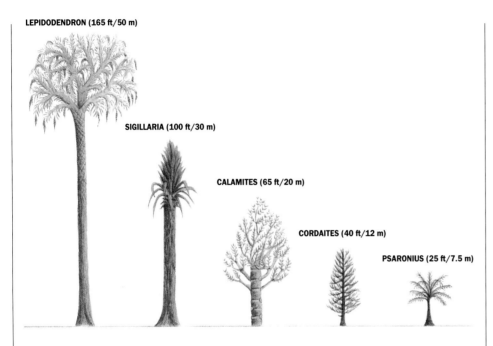

LEPIDODENDRON (165 ft/50 m)

SIGILLARIA (100 ft/30 m)

CALAMITES (65 ft/20 m)

CORDAITES (40 ft/12 m)

PSARONIUS (25 ft/7.5 m)

TREE HEIGHT *Carboniferous trees ranged in height from giant club mosses like* Lepidodendron *and* Sigillaria, *horsetails like* Calamites, *to* Cordaites *and ferns such as* Psaronius.

and survive to this day in the tropics. The swamps were also home to a group of tree-sized plants, generally known as *Cordaites*, which originated during the Carboniferous era. Ancestors of our conifers, these plants had woody trunks and branches, which made them the first true trees and enabled them to grow to heights of 100 ft (30 m) and more. These trees lasted into the Permian era, when they were probably pushed out of their habitats by the true conifers. The true conifers have the longest fossil history of any living seed plants; the first fossils are found in the Late Carboniferous era. Their ability to live in drier conditions, above the water level of the coal swamps, was a significant innovation. Fossil evidence suggests that they had branching patterns and leaves similar to the Norfolk Island Pine, *Araucaria*, of today.

Our knowledge of all these ancient plants has been painstakingly put together by plant scientists from all over the world over the past 200 years or so. And yet the fossils studied by these palaeobotanists are no more than fragments of plants, found in sites probably far from where the parent plants originally lived. This is because rivers have transported them hundreds of miles from their place of origin to the site where they were eventually deposited and fossilised

in layers of sediment. The same processes take place today. On most of the land surface, plant material decays, releasing organic carbon into the soil and carbon dioxide into the atmosphere. It is only when leaves, branches, torn-up trunks and roots are buried quickly in the sediment that there is much chance of preservation.

WHERE FOSSILS ARE FORMED

This only happens when a river, especially one in flood, erodes its banks, tears up vegetation from the valley bottom, and carries the debris downstream before dumping it on the flood plain, where it is buried in deposits of sand and mud. Some of the lighter and more buoyant debris, such as leaves, fruit and pollen, is carried hundreds of miles downriver to the estuary, delta or

FERN FRONDS *Carboniferous true ferns, like* Neuropteria *(below left), required moist habitats, while seed ferns, like* Sphenopteridium *(right), tolerated drier conditions.*

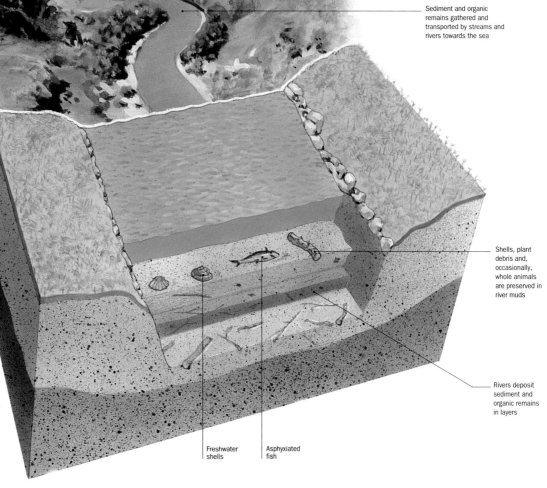

Dead dragonfly and fern frond before burial in the mud

BURYING THE PAST *Fragile plants and insects disintegrate after death and decay. For them to be preserved, they need to be buried rapidly in fine sediment such as river mud.*

Weathering and erosion of hills into valleys

Sediment and organic remains gathered and transported by streams and rivers towards the sea

Shells, plant debris and, occasionally, whole animals are preserved in river muds

Rivers deposit sediment and organic remains in layers

A MOVING DEATH
Migrating river channels can carry the bodies of dead creatures far away from the point of their demise. Hence the presence of fossil creatures in a particular rock stratum is not necessarily an indication that they all existed in the same geographical region.

Freshwater shells

Asphyxiated fish

even out to sea, where it may be buried in offshore marine sediments.

As a result, the scientist is often confronted with a very mixed collection of fossil debris, in the form of stems, leaves or fruit that may or may not be from the same plant. The problem with trying to piece the jigsaw together is that there is no picture to show what the final assembly should look like. Nor do the pieces necessarily interlock in any helpful way.

What palaeobotanists have been able to

A FOSSIL FOREST

Lying within the suburbs of the city of Glasgow in Scotland, Victoria Park is the site of the most famous fossil forest in the world. In 1887, 11 fossilised lycopod tree stumps were found clustered together with their root systems, where they had originally grown some 325 million years ago. These stumps are the oldest preserved example of part of an Early Carboniferous forest of giant club mosses. Rooted in mudstones, a form of fossil soil, the closely spaced 'trees' indicate an original density in this ancient equatorial forest of around 11 250 per sq mile (4344 per km²).

establish is that the coal seams of the Late Carboniferous era were formed from the peaty deposits of inland swamps and coastal deltas around 320 million years ago. Analysis of the fossil plant debris that made up these peats shows that the original swamps were dominated by giant club mosses whose

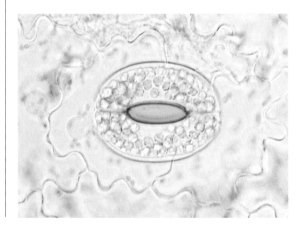

TREE-SIZED FERNS Although Dickinsonia-type tree ferns only date back to Jurassic times, tree-sized Carboniferous ferns would have looked very much like them.

remains are preserved in situ. The other major groups of plants grew elsewhere. Just as the main sites of real plant diversity today are beside the world's tropical rivers, the horsetails, ferns, seed ferns and tree-like plants of the Carboniferous era flourished beside rivers, which drained into the swamps and deltas. Unfortunately, these river banks themselves are rarely preserved as deposits, because they have been constantly undercut and eroded by the river as it shifts its course and meanders its way to the sea. The remains of the plants that once grew on the banks are, therefore, generally only preserved as fossils where they were deposited by the river, miles from their site of origin.

There is very little fossil evidence, however, of the kinds of plants that might have lived in the drier uplands, beyond the wetlands of the swamps, coastal deltas and river banks of the Late Carboniferous period.

FOSSIL AIR?

Nevertheless, analysis of Carboniferous plant fossils has given scientists a remarkable measure of the dramatic increase in vegetation more than 300 million years ago. The 'greening' of the land, which took place between the Devonian and the Carboniferous eras, was so extensive and luxuriant that the composition of the atmosphere changed. For the first time in

BREATHING SPACE Leaf cells, called stomata, allow plants to 'breathe' and exchange atmospheric gases, such as oxygen and carbon dioxide.

Earth's history, the level of carbon dioxide, the main 'greenhouse gas' in the atmosphere, fell dramatically as plants absorbed it – and the proportion of oxygen rose.

Palaeobotanists have developed an ingenious means of measuring the composition of past atmospheres. During photosynthesis, plants absorb carbon dioxide, CO_2, from the atmosphere through pores, or stomata, on their leaf surfaces; as scientists have discovered, they respond to any changes in atmospheric composition by altering the number and density of these stomata over time. To regulate the rate at which the plants assimilate CO_2, the stomata are less densely packed when there are high levels of CO_2 in the atmosphere and more highly packed when the levels are low.

By measuring the density of the stomata on fossil plants, scientists can monitor fluctuations in CO_2 over the last 400 million years or so. Measurements from Devonian plants, some 395 million years old, indicate levels of CO_2 that are between 10 and 20 times higher than those of today; while measurements from 300-million-year-old Permo-Carboniferous 'coal forest' plants record levels that are only twice today's.

Plants are the organisms that really

THE ROOTS OF A COAL FOREST

Short columns of pitted stone are some of the most common fossils in Carboniferous coal-mining areas all over the world. Museum specimens date back to the 18th century, and they were first named *Stigmaria* by the French naturalist Adolph-Théodore Brongniart in 1828. But it was not until the exploitation of the coalfields later in the century – and particularly those in the north of England – that scientists discovered what they actually were.

During the construction of the Manchester to Bolton railway in 1839, workers discovered a series of large fossil tree bases with spreading roots. The plant remains consisted of sandstone casts – a means of preservation which the

BACK TO THE ROOTS
These branching Stigmaria *roots were found in Murgatroyd's Quarry near Bradford, England, in 1886.*

naturalist John Eddowes Bowman explained by comparison with the way trees rot in tropical forests today. The woody centres of the stems and roots can decay within a year, leaving a hollow trunk of outer bark, which is then filled with sediment when the forest floor is flooded. Bowman also proved that the trees had grown where they were found, by showing that the forked root systems extended down into the coal below.

Plaster casts were made of the best examples, and can still be seen in Manchester Museum today. Subsequent similar finds, made in the region in 1844, demonstrated that *Stigmaria* roots are related to a tree that had independently been named as *Sigillaria* – a connection confirmed in 1848 by discoveries in the coalfields of Nova Scotia. Another Manchester naturalist, William Crawford Williamson, who was the first professor of geology at

Manchester University, described the detailed anatomy of these giant club mosses in 1887.

The year before publication, Williamson received news that another stump had been found in Murgatroyd's Quarry, near Bradford. Williamson managed to secure the specimen for Manchester Museum, and the publicity attending the recovery of the 5 ton specimen

PETRIFIED TRUNK *Sandstone filled the hollow trunk of a Carboniferous lycopsid tree, while it stood in a coal forest over 300 million years ago.*

guaranteed its international fame. The 310-million-year-old tree still has its *Stigmaria* roots extending for many yards on either side of its 5 ft (1.5 m) wide trunk.

EARLY REPTILE *The skeleton of* Silvanerpeton *suggests that the 12 in (30 cm) long Carboniferous reptile was aquatic, but it may have been only a juvenile.*

ruled the Carboniferous era. However, the gradual movement of plants onto drier land in the Upper Carboniferous era opened up new habitats that would require of any vertebrates that grasped the opportunity a far greater independence from water. In particular, these creatures needed to develop an egg with a tough membrane – the cleidoic egg – that allows the developing embryo to breathe and excrete while in its egg capsule, without having to obtain oxygen from water. As a result, the egg can be laid on land.

THE FIRST REPTILES

Today, the reptiles that evolved the ability to lay such an egg are easily distinguished from the amphibians, because they have evolved separately for 300 million years. But their characteristic features – the possession of dry, scaly skins and the ability to lay cleidoic

PRIVATE POND *Scaly-skinned reptiles, ancestors of this green mamba snake, were the first vertebrates to produce shelled eggs and to breed on land.*

eggs – are not readily fossilised. So how can the first reptiles be identified?

Fortunately, there are various skeletal innovations that help scientists to distinguish amphibians from reptiles. The first generally accepted fossil reptiles are around 300 million years old and were originally found, more than 150 years ago, in a Mid Carboniferous coalfield in Nova Scotia. During the mid-19th century, the coalfields of the world's industrial nations were being explored by geologists for the first time, in order to keep pace with the demand for coal. And so it was that, in the 1840s, near Joggins, Nova Scotia, the fossilised trunks and roots of a number of large tree-like club mosses, or lycopods, were found in their original growth positions.

Known as *Sigillaria*, they were still standing upright after 300 million years, surrounded by layers of sandstone, mudstone and coal. By 1852, over 30 of these fossil stumps had been unearthed and, much to the astonishment of geologists, some were found to contain bones. Altogether, there were the skeletons of several hundred small amphibians and several reptile species, as well as the remains of snails and millipedes. Among the reptiles, there were two lizard-like creatures, with slender bodies

about 8 in (20 cm) long from the snout to the tip of the tail. Unlike most amphibians, the head is relatively small and the skull lightly built, with small conical teeth. Called *Hylonomus* and *Palaeothyris*, they are thought to have fed on insects and other small land-living arthropods, such as millipedes. A key innovation was the development of extra muscles to strengthen the biting power of the jaws. The combined effect of sharp, spiky teeth and these muscles would have been quite enough to crunch through the tough exoskeleton of a millipede.

The *Sigillaria* 'trees' grew to heights of up to 100 ft (30 m) in the lakeside marshes of ancient Nova Scotia. From time to time, the lakes had flooded and deposited layers of sand and mud on top of the swampy peat that surrounded the base of the trees. When the trees eventually died, the upper part of the stems, branches and leaf canopy rotted and fell away, while the lower portions were protected and held upright by the surrounding sediment. Even so, the plant tissues of the lower portions also rotted, with the boles becoming hollowed out, in much the same way as happens to trees today. At first, it was thought that the animals lived within the hollows, but more recent work has shown that the hollows acted like bottle traps. Some of the animals that fell into them could not then climb out, and only survived so long as they could feed on any invertebrates, such as snails and millipedes, that had inadvertently wandered into the same trap.

It was from such beginnings – captured so perfectly by this chance window on primitive life – that the reptiles evolved and soon reached out to occupy terrains that were not available to the amphibians.

LIFE'S MIDDLE AGE

2

RISE AND FALL *For 200 million years, the dinosaurs made an indelible mark on Earth.*

IN THAT TIME OF LIFE BETWEEN JUVENILITY AND MATURITY, THE MIDDLE AGE REFLECTED A PERIOD OF CONSOLIDATION FOR THE ALL-POWERFUL RULERS OF PLANET EARTH. DINOSAURS HAD ALREADY ACHIEVED WORLD DOMINATION OVER OTHER LAND CREATURES, SAFE IN THE TYRANNY OF THEIR MONSTROUS SIZE AND FEROCITY. MEANWHILE, THEIR DISTANT REPTILIAN RELATIVES HAD SET UP THEIR OWN DICTATORSHIP OF THE SEAS AND SKIES. BUT EVEN AS THE REPTILES HELD THE WORLD IN THEIR SAVAGE GRIP, THERE WAS A QUIET REVOLUTION GATHERING MOMENTUM AT GROUND LEVEL. IN ONE FELL SWOOP THERE WOULD BE TOTAL WIPE-OUT. BUT DID THE DINOSAURS FALL OR WERE THEY PUSHED?

RULED OUT *Large flying reptiles dominated the skies.*

HOW THE REPTILES CAME TO RULE

The kingdom of terrible – and not so terrible – lizards

spanned creatures small enough to fit into a jacket pocket

and giants that were almost the size of a blue whale.

All had a body plan that was set for world domination.

The image of the dinosaurs as the 'ruling reptiles' of the Mesozoic era is so pervasive nowadays that the long history of their ancestors and all other reptilian creatures – from their Late Carboniferous beginnings right through to the present – tends to be overshadowed. And yet all the other reptiles are just as interesting and as important in the history of life before man.

Reptiles are still very much in evidence today, with over 6000 species alive – and they still easily outnumber the mammals (at over 4000 living species). Most of these reptiles are lizards, which are an incredibly successful group. Not only do they have a long history but they also have a body plan that has proved to be enormously adaptable for life in many different habitats – from the driest of deserts to rivers and seas and even to gliding through the air.

ERAS OF EVOLUTION

The Late Carboniferous, Permian and Triassic phases of geological time, stretching from around 300 million to 205 million years ago, were important eras in the early evolution of the reptiles. Towards the end of the Carboniferous era, the continents and subcontinents known today as South America, southern Africa, Antarctica, India and Australia were clustered around the South Pole, forming a landmass known as Gondwanaland. As global temperatures fell, a southern ice cap and associated glaciation spread over the landmass, cutting back the vast coal-producing forests and swamps in the process.

During the Permian phase, underlying geological changes caused this cluster of southern continents and subcontinents to move north. As they moved closer to the Equator, and so warmed up, the ice cap diminished and then disappeared. Eventually, in Mid Triassic times, about 220 million years ago, the southern Gondwanan conglomeration met up with the northern continents of North America and Eurasia to form the supercontinent of Pangaea, meaning 'whole earth'.

This massive landmass, which covered approximately two-fifths of the surface of the globe, was surrounded by a super ocean sometimes referred to as Panthalassia. The interiors of the continents were largely hot and semiarid, and extensive salt deposits developed in North America and Europe during the Triassic period as inland seas and

A MODERN CLASSIC *The living tuatara, from New Zealand, epitomises 'lizardness' with its limbs sticking out on either side and long, muscular tail.*

RELATIVE VALUES *For 165 million years, the dinosaur reptiles dominated the landscapes of the Earth – the 800 species we know of are only a fraction of what existed. Family trees attempt to show how they were related and evolved.*

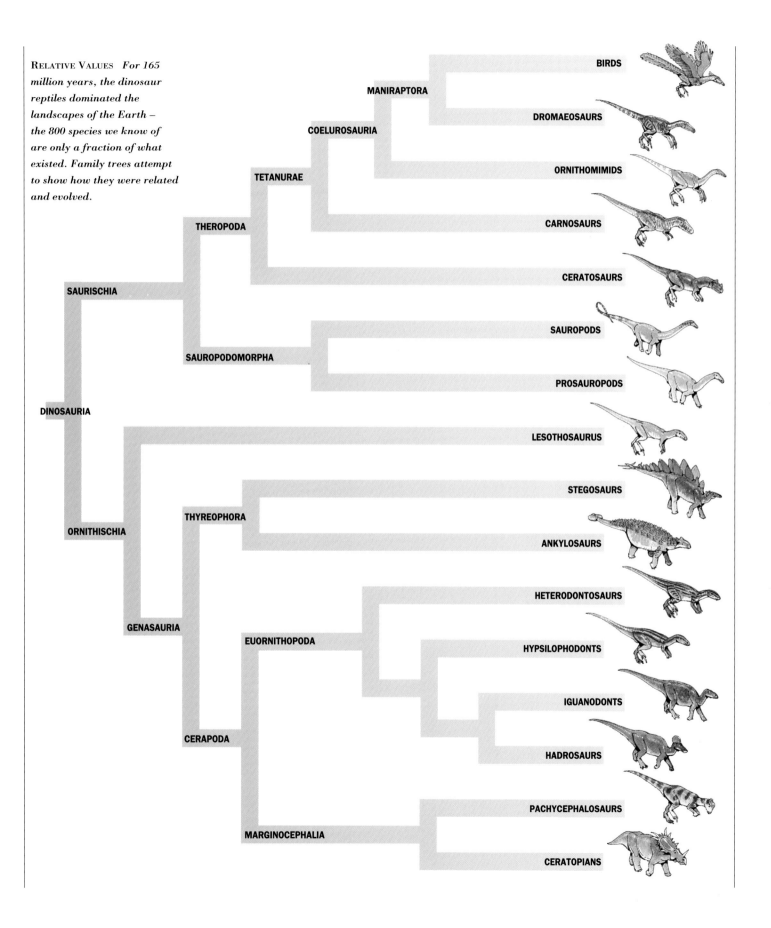

BIRDS

MANIRAPTORA

DROMAEOSAURS

COELUROSAURIA

ORNITHOMIMIDS

TETANURAE

CARNOSAURS

THEROPODA

CERATOSAURS

SAURISCHIA

SAUROPODS

SAUROPODOMORPHA

PROSAUROPODS

DINOSAURIA

LESOTHOSAURUS

STEGOSAURS

THYREOPHORA

ANKYLOSAURS

ORNITHISCHIA

HETERODONTOSAURS

GENASAURIA

EUORNITHOPODA

HYPSILOPHODONTS

IGUANODONTS

CERAPODA

HADROSAURS

PACHYCEPHALOSAURS

MARGINOCEPHALIA

CERATOPIANS

REVELATIONS OF THE DEEP: THE SOLNHOFEN LAGOON

Stone has been excavated from the quarries in the Jurassic sedimentary rock of southern Bavaria, Germany, for centuries. Valued for its ability to split into large, smooth-surfaced flags of varying thickness, the stone is cut from fine-grained limestones laced through with occasional thin flat laminae (rock beds). These allow the layers of rock to separate cleanly.

The calcareous – or chalky – mud that makes up each flat bedding plane was originally the sediment of a shallow lagoon, and as such was an excellent preserving medium for fossils of the future. Evidence from both the sediment and the fossils suggests that the large lagoon was dotted with low-lying islands, protected by a reef from the Tethys ocean lying to the south (whose sediments now form the Alps).

The fossils found in the rocks are remarkably diverse, but are dominated by crustaceans and fish. Shellfish that would have lived on the sediment surface are poorly represented, which suggests that the bottom water and sediments of the lake were probably low in oxygen and therefore toxic for many creatures. The lack of oxygen was probably caused by high evaporation rates – it was hot in this part of the world at the time – and poor circulation of the bottom waters. However, as a result, many of the fossils are extremely well preserved.

Most of the non-marine fossils of Solnhofen are the remains of flying animals, such as insects and pterosaurs, although at least seven specimens of *Archaeopteryx* have been found by quarry workers over the years. Without the unusual preservational conditions of the carbonate muds of the Solnhofen lagoon, which preserved impressions

PAST MASTERS *Originally a source of limestone slabs for lithographic printing (far left), the Solnhofen quarries are now worked more for their beautifully preserved fossils (above).*

of the feathers, this unique Jurassic animal might never have been recognised as a bird. It would have been classified as just another small theropod dinosaur.

---

drainage basins dried up. The climatic changes also had a drastic effect on the vegetation. Seed-bearing plants and conifers of a more modern type flourished in the north, while a group of distinctive plants related to *Glossopteris* developed in the south – replacing the club mosses and horsetails of the Carboniferous vegetation.

EARLY RECORDS OF THE REPTILES

Fortunately for fossil-gatherers today, the fact that sea levels remained low during these times of change meant that significant volumes of terrestrial sediment built up on the continents.

Understanding the evolutionary history of the reptiles relies in part on the quality of fossils that have been recovered so far, and how well they are preserved depends on the type of sedimentary rock they were deposited in. Although the corpses of some land-living animals were eventually covered over on dry land, where the creatures died, many were transported by rivers and became submerged in sediments along the coastline. As fossils are often best preserved in marine and coastal sediment, it is lucky for collectors that this is the case. The Jurassic lagoonal carbonate muds of Solnhofen in southern Germany and the Cretaceous Wealden deposits of southern England have turned up some remarkable finds. Indeed, some of the earliest dinosaur discoveries were made in sedimentary rocks in the south of England, originally deposited when the area was a low-lying coastal plain.

Inevitably, however, interpetation of fossil evidence is constantly changing. The recognition of fossil reptiles depends upon finding well-preserved skeletal features and relating them to their nearest living equivalent. This is particularly difficult with many of the early extinct reptile-like groups as they have no living near-relatives. Most finds of Late Carboniferous and Early Permian tetrapods (four-legged vertebrates) have come from the Northern Hemisphere, perhaps because this warmer region was more habitable for them.

THE SIGNIFICANCE OF THE REPTILES

Reptiles can be divided conveniently into three categories – creatures of the land, the sea and the air – and it was this adaptability to different conditions that allowed them to evolve into such a success story. One of their most fundamental and significant attributes is internal fertilisation and subsequent early growth and development of the embryo within a protective (amniotic)

THE SURVIVORS *Modern crocodiles – and relatives such as alligators – are archosaur reptiles whose ancestry predates the dinosaurs.*

membrane. This important characteristic differentiates the four-legged (tetrapod) vertebrates into those that do possess it (known as the amniotes) and those that do not (the anamniotes). The amniote group includes birds and mammals as well as reptiles.

In evolutionary terms, amniote reptiles were able to establish themselves so well because of their ability to lay shelled eggs. Unlike anamniote amphibians, they could invade dry land and not have to return to water to breed.

THE REPTILES RETURN TO THE WATER

Although the evolution of the reptiles was a fundamental part of the colonisation of the land, fossil discoveries show that they did not turn their backs on an aquatic existence forever. No sooner had the early vertebrates become independent of water than some promptly did a U-turn and took up the aquatic habit again.

This is clearly shown by the fossil remains of a 12 in (30 cm) lizard found in Kansas. This Late Carboniferous fossil, named *Spinoaequalis*, which dates from 300 million years ago, gives a clear indication that the lizard's tail was adapted for swimming. It would seem that this single fossil represents the first amphibious reptiles to hedge their bets by not becoming fully committed one way or the other to the

land or sea. What the fossil evidence also indicates in a roundabout way is that the ability to lay shelled eggs may not have been as liberating as was previously thought.

Spinoaequalis belonged to the diapsid group of reptiles, which was not only the first to return to the water but also continued to do so throughout geological time. The diapsids included dinosaurs as well as the majority of surviving reptiles, such as modern lizards, crocodiles and snakes, all of which have well known aquatic representatives. It is likely that some aspect of their biology gave them a selective advantage for taking the plunge into water.

Although the Kansas fossil lizard is not complete, many of its tiny, delicate bones were still joined – and it is unlikely that the corpse had been moved any great distance before it was buried. The Upper Carboniferous marine muds in which the find was made also contain many beautifully

preserved marine fish, and it would seem from these fossil associates that the lizard was capable of swimming in the sea.

As the first reptiles are thought to have emerged about 350 million years ago, it seems it was not long – in evolutionary terms – before the reptiles began to diversify. When the fossil *Spinoaequalis* is compared with the only other water-living fossil reptile of this kind, *Hovasaurus* – which dates from 255 million years ago and is considerably younger – at first there seem to be startling similarities. Like *Hovasaurus*, the tail vertebrae of *Spinoaequalis* had unusually long upper and lower spines for muscle attachment and no transverse projections. The width of the flattened tail was around a quarter of its length, and its articulated tail bones allowed for considerable sideways movement, a configuration which is typically associated with swimming.

Where the two fossils differ is in the development of the limbs. Fully aquatic reptiles such as *Hovasaurus* and other animals often show modifications in the limbs

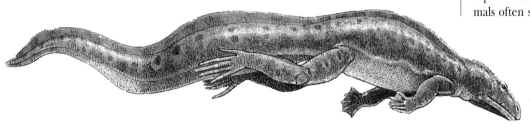

RETURN TO WATER *Although reptiles are essentially adapted for life on land, some, such as Hovasaurus, preferred an aquatic existence.*

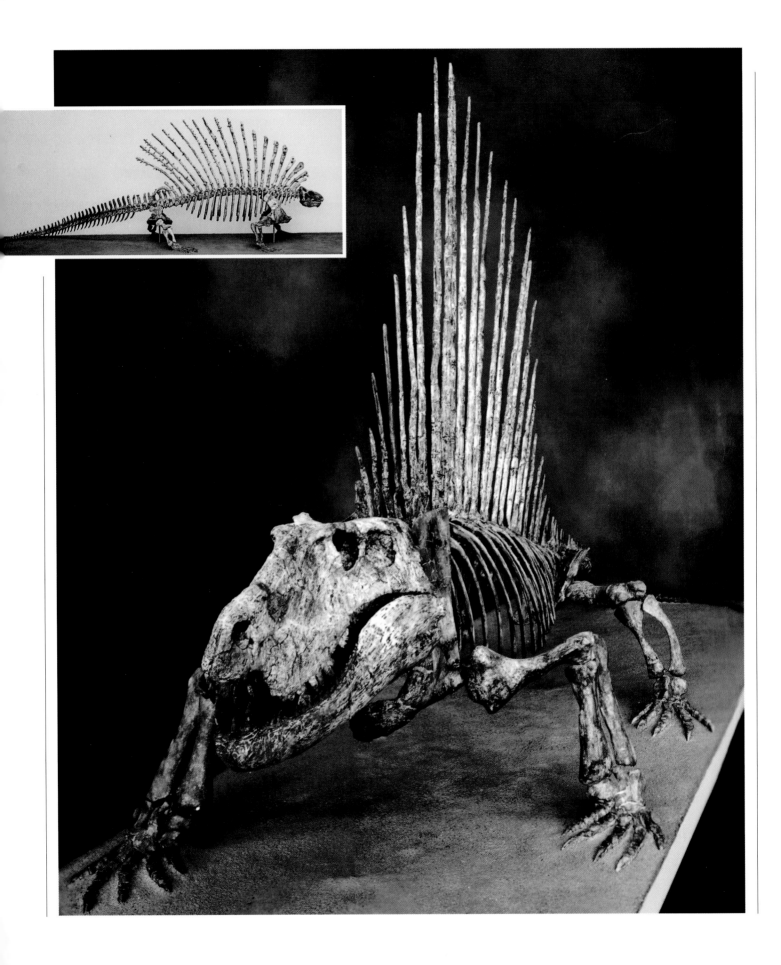

for swimming. The limbs of *Spinoaequalis* are relatively thin and long, however, and are generally adapted for walking on land.

THE SAIL-FINNED LIZARDS

Some reptiles may have been developing in the water, but the main progress in their evolution took place on land. Pelycosaurs or sail-finned synapsids first appeared in the middle of the Upper Carboniferous era, and were soon diversifying in the hot, semi-arid deserts of Early Permian times. The rocks formed at that time are today represented by sandstone in areas such as Texas.

The pelycosaurs were members of a major grouping of reptiles known as the synapsids, which had particular skeletal characteristics that allowed them to be distinguished from other major groups (such as the diapsids). The synapsids are particularly important because successive groups become more mammal-like through a gradual but continuous acquisition of mammalian characteristics. In the pelycosaurs, this was apparent in the beginnings of variation in tooth form.

The most famous of this group are the lizard-like *Edaphosaurus*, which measured roughly 8 ft (2.4 m), and *Dimetrodon*, which reached up to 9 ft (2.7 m) long. Both had limbs sticking out sideways from the body, and these right-angled joints resulted in a sinuous walking movement similar to that of the earlier tetrapods. Most striking of all was a dorsal fin that stuck up from their backbones. This was made from enormous elongated spines, which were then covered in skin to form a sail-like structure.

Both reptiles show significant advances in their form over earlier groups, and had developed a means for more efficient feeding. The small-skulled *Edaphosaurus* was able to predigest plant food, an advance on previous plant eaters thanks to its more sophisticated jaws and palate. These held peg-like teeth which were able to crush plants and break down the tough cellulose of the cell walls before being swallowed.

Dimetrodon had a much larger skull and the characteristically longer and sharper teeth of a carnivore – but with some differentiation in size. The larger, dagger-like teeth were for seizing and killing prey, and the smaller teeth were for ripping it to pieces. The earliest carnivores were unable to tear their prey, and so had to swallow it

whole, just as some living reptiles – such as crocodiles and snakes – still do today. For such cold-blooded animals, this primitive arrangement is not much of a drawback as their energy requirements are relatively low and they do not need to catch and eat prey every day. Carnivorous mammals, who need a lot of energy to maintain their higher body temperatures, have to spend more time catching more prey and eating it. Like modern reptiles, pelycosaurs were probably cold-blooded ectotherms, whose metabolism did not heat their bodies fully and who

had to rely on external sources to become completely active. Although it is unclear what the exact function of the pelycosaur sail-fin was, there is evidence that the web of skin that stretched between the bony sail supports and the supports themselves was well supplied with blood vessels. This suggests that the sail regulated the animal's temperature and helped it to either gain or lose heat when necessary.

Without a heat-exchanging sail such as this, it might have taken a large 551 lb (250 kg) *Dimetrodon* as long as 12 hours of basking in the sun to raise its body temperature from 25℃ to 30℃ (77℉ to 86℉). With

PERMIAN TREASURE CHEST
So far, the Karroo desert in southern Africa has yielded more than 200 different kinds of Permian fossil reptiles.

the extra surface area provided by a sail, it would take the animal only two hours to gain the same increase in temperature. As a predator, this would have provided a considerable advantage over more torpid prey. Although the sail-fin developed at least three times – independently, and in

BUTT IN *The behaviour of modern animals in combat helps the understanding of extinct creatures with similar physical attributes.*

herbivorous and carnivorous pelycosaurs – not all of these creatures possessed it. This implies that it was not an essential structure for the survival of the species.

THE THERAPSIDS

By Mid to Late Permian times, there was a considerable variety of life on land. Ecological relationships were becoming more complex as the various species diversified to fill specific habitats, and the growing numbers of reptiles included a range of carnivores and herbivores similar to the pelycosaurs. These fascinating and often bizarre-looking animals are collectively referred to as the therapsids, and deserve to be better known. For too long, they have been overshadowed by the image of the dinosaurs.

Both the Perm region of Russia and the Karroo desert in southern Africa have produced important fossil records of land

vertebrates, with over 200 different kinds of fossil reptile recovered from the Mid to Late Permian sediments of the Karroo. One of the best known of these is a therapsid called *Moschops*, an extraordinary beast, which grew to 15 ft (4.6 m) and had a massively built body, including a strong ribcage and heavy limbs. Its hands, feet and skull were relatively small, and its innovative legs could be tucked in under the body and pelvis. The forelimbs still stuck out sideways, as with earlier reptiles.

It is thought that the head was held relatively high, supported by a thick muscular neck and deep shoulders leading into a sloping back, quite unlike the way most modern lizards are built. These proportions and the frame are typical of a

plant-eating animal, however, and *Moschops* is an early example, in evolutionary terms, of one strategy for survival by plant-eaters.

Herbivores have to consume a considerable volume of vegetation to gain sufficient energy because of the low protein levels in most plants. As the chemical process of digesting plant material is relatively slow, the stomach has to be large enough to accommodate all of the food – and this needs to be supported by a substantial skeleton with muscles to move it. The drawback for such creatures is that this bulk would slow them down if they needed to outrun an enemy, and their teeth – designed to cope with tough leaves – would be of little use as a defence against a carnivorous predator.

An alternative defensive strategy developed by *Moschops* was to turn its bulk to good advantage. Its tactic was to stand its ground and face up to its enemies rather than running away (so exposing its unprotected back and hindquarters). The belly was protected by the heavy ribcage, and its muscular limbs suggest that, although *Moschops* was not a fast runner, it was probably fairly nimble. Its key weapon, however, seems to have been its skull, which had extraordinarily thick roof bones that were up to 4 in (10 cm) thick.

HEAD FIRST *A thick skull supported on muscular shoulders suggests that the plant-eating* Moschops *faced up to its enemies.*

To make sense of such structures in extinct creatures is only possible in comparison with living creatures with similar physical attributes. Various wild sheep and goats alive today have skull bones of this kind, which they use for head-butting in defence

MODERN HEAD-BANGERS

North American bighorn sheep are living examples of animals that habitually use their skulls as weapons in head-to-head fighting. When American biologists Caroline Jaslow and Andrew Biewener analysed the impact force of these head-banging males in combat, they discovered it could be as high as 3400 newtons – 60 times greater than the force required to break a human skull. These living mammals largely absorb the potentially mortal blows thanks to the complex sutures which hold the skull bones together. The sutures act as shock absorbers, flexing as each blow is received.

of territory, for the protection of young, and in competition between adult males for dominance and access to females. By lunging forward with all its considerable weight and using its thickened skull as a battering ram to head-butt an attacker, *Moschops* would have made a formidable opponent. Any damage to the brain and spinal cord would have been avoided by safely transmitting the impact down its thick neck bones and into the powerful shoulder girdle.

MOVING TOWARDS MAMMALS
Moschops is a member of just one of the groups of therapsid reptiles that dominated the landscapes of the Late Permian and Triassic periods. Another important group were the dicynodonts, some of which were up to 9 ft (2.7 m) long. These were the main vertebrate herbivores at that time, and their abundance suggests that originally they fulfilled a similar ecological role, at the bottom of the vertebrate food chain, to large modern grazing mammals such as buffalo and rhinoceros. The considerable evidence of this group found in southern Africa (where their fossils have sometimes

accounted for up to 90 per cent of animal remains) shows modifications in their skulls and teeth. These would have helped them deal more efficiently with their vegetarian diet, and the success of the dicynodonts is most likely the result of their evolution of sophisticated jaws and teeth.

Most earlier reptiles had jaws that worked in a similar fashion to a pair of scissors, which is fine for cutting foliage off branches, for example, but not so suitable for grinding up coarse plant material. To do this efficiently, some sort of horizontal motion is required – with opposing sets of teeth rubbing against one another. Such a motion needs quite complex jaw articulation and musculature.

One small animal that seemed to resolve this problem was *Emydops*, which had a 2 in (5 cm) long skull, a short, beaked snout and a reduced number of teeth compared to other early reptiles. The sharp beak may have had a horny covering, similar to the shell of a turtle, and would have been effective in tearing leaves away from plants. Sharp canines were retained

in the upper jaw only, then there was a gap in the cheek region. A battery of small, tightly packed, 'post-canine' teeth could be found set far back in the mouth, where they formed a grinding surface similar to that created by the molar teeth of mammals.

Analysis of the wear patterns shows that the jaws of *Emydops* were exceedingly mobile compared to those of earlier species. When eating, the lower jaw slid forward to make the initial bite, then the plant food

was dragged back to the post-canines with the tongue. As the jaw was retracted, the post-canines would grind up the plant food. It is because of these mammal-type traits that the dicynodonts are often referred to as the 'mammal-like' reptiles.

THE GORGONOPSIDS
At the pinnacle of the food chain were the top carnivores, which, during the Late Permian period, came from another therapsid group called the gorgonopsids.

By comparison with other creatures at the time, gorgonopsids such as *Lycaenops* and *Arctognathus* were small – measuring around 3 ft (1 m) long. But with their muscular and agile body build, powerful limbs for running and a long tail for balance,

PURE PREDATOR *Strong jaws and differentiated teeth with tusk-like canines are characteristic of mammal-like reptiles such as* Cynogathus.

they were the Late Permian equivalents of the sabre-toothed cats of the Pleistocene era. Gorgonopsids had relatively large skulls with enlarged, dagger-shaped canine teeth and a jaw articulation that could open through 90 degrees. Once they had caught their prey, they would have stabbed it in the neck and disabled it by strangulation or by severing the main nerve cord. Bite-sized

continued on page 66

REPTILE REIGN *Permian times, more than 250 million years ago, saw the first well-established land communities of large vertebrates, in places such as the Karroo area of South Africa. Amphibians dominated fresh waters and their environs, and egg-laying reptiles ruled farther inland. Such reptiles evolved long before the dinosaurs and included animals with mammalian features (for example, differentiated teeth). These mammal-like reptiles included herbivores – such as Moschops, Pareiasaurus and Dicynodon – and carnivores such as Lycaenops, Youngina and Titanosuchus.*

TITANOSUCHUS

LYCAENOPS

DICYNODON

YOUNGINA

OSCHOPS

PAREIASAURUS

SECRET SUCCESS *A history of burrowing has resulted in the unfamiliar caecilians, such as* Gymnophis, *losing limbs and sight.*

amphibians (collectively known as the lissamphibia) have been hard to establish because so few fossils charting their development from their Palaeozoic ancestors have been found. It is clear that the three lissamphibian groups – represented by frogs, salamanders and caecilians – were already established by the Late Triassic era, and must have originated in Early Triassic times.

The amphibians of today – which include frogs, salamanders, toads, newts and the remarkable (not quite so familar) worm or snake-like caecilians – are radically different in size and demeanour from their numerous Palaeozoic ancestors.

Frogs have the oldest fossil representatives, some dating back to the Late Triassic era 210 million years ago. These show that the body plans of adult frogs today are radically different from their ancestral amphibian form. Then, they were much more like salamanders and late-stage tadpoles, with well-developed tails and four limbs of more or less equal size.

The second group, the salamanders, appears to be relatively unchanged from the earliest representatives of their group to be recovered, which are Mid Jurassic in age.

The third group are the worm-like caecilians, which today can measure up to 3 ft 3 in (1 m) in length. Living caecilians are much more common than might be thought, especially in the tropics, with 162 different living species. These extraordinary, secretive creatures are well-adapted to a burrowing mode of life now, and have

limbless, elongated bodies which are made up of between 82 and 285 vertebrae. This compares to the 12 or so vertebrae found in short-bodied frogs. The caecilian's worm-like form is emphasised by the folds of skin which form a series of rings around its body. Caecilians also have a compact skull with a well-developed bony braincase and strong jaws. Their eyes are small and often covered in skin, and some species have small, unique, extending 'tentacles' at the angle of their jaws, which serve as sensitive feelers and compensate for the loss of sight.

These tropical amphibian 'worms' are active carnivores that can sense and seek out their prey underground, armed with relatively large mouths filled with two rows of numerous small, sharp teeth. Because of their underground life, some of them have become independent of water, even for reproduction and, unlike other living amphibians, they are viviparous – the young develop internally and are fed by special secretory glands. As a result of all these modifications and, until now, the lack of useful fossil record, it has been difficult to determine the evolutionary relationships of caecilians – to other living lissamphibian

LIVING REMINDERS
Salamanders are a relatively recent group. They resemble many of the extinct Palaeozoic amphibians.

relatives or to their Palaeozoic amphibian ancestors. However, the fact that they are found throughout the tropics worldwide suggests that they have an ancient history, predating the final breakup of the Pangaean supercontinent, and that they are Gondwanan relics.

The discovery of a 200-million-year-old fossil caecilian in Arizona in 1992 has gone some way to unravelling the history of these limbless amphibians. The fossil, *Eocaecilia micropodia*, displays a uniquely caecilian sensory organ and specialised jaw, yet shares features with salamanders and one of the numerous Palaeozoic groups of amphibians, the microsaurs. (This ancient group became extinct in Early Permian times, 280 million years ago.) Most important of all, the fossil has legs but unusually small feet.

Eocaecilia clearly lived a different type of life from its modern relatives, despite being similar in form and size. It still had limbs, although none had more than three toes, and its eyes were not reduced, suggesting that the fossil caecilian had not yet adopted a burrowing mode of life.

THE FORERUNNERS OF CROCODILES AND DINOSAURS

With the onset of the Triassic era, the polar ice caps disappeared and the world's climate improved. If anything, there was a gradual shift from warm, moist weather to more hot

CRYSTAL PALACE — THE WORLD'S FIRST THEME PARK

One of the greatest boosts to the development of science and the study of fossils was provided by the opening, on June 10, 1854, of the rebuilt Crystal Palace by Queen Victoria at Sydenham in south London. Not surprisingly, over 40 000 people turned up to see the Queen perform the opening ceremony.

The grounds around Joseph Paxton's magnificent glass and iron 'Palace of the People' – created for the Great Exhibition in Hyde Park

MODEL STANDPOINT *The 19th-century view of dinosaurs at Crystal Palace was based on the four-footed stance of living lizards such as* **Tuatara.**

three years before – had been developed into the first educational theme park in the world. A huge display of over 30 lifesize prehistoric 'monsters' had been built and placed in natural settings of rocks, plants and lakes. They had an immediate impact on popular imagination and a wave of 'dinomania' swept the country.

The display was the brainchild of Sir Richard Owen – with the support of the Prince Consort, Albert – who saw an opportunity to promote his own theories of animal relationships against the growing interest in the evolutionary ideas of Charles Darwin.

An obscure artist, Benjamin Waterhouse Hawkins, was chosen

'to exhibit, restored in form and bulk, as when they lived, the most remarkable and characteristic of the extinct animals and plants of each stratum'. Three islands were laid

DINOSAUR DINERS *The inside of an* **Iguanodon** *was the venue for Hawkins's New Year's Eve banquet at Crystal Palace, 1853.*

out to represent the three main divisions of 'fossil' time – the Primary (Palaeozoic), Secondary (Mesozoic) and Tertiary (Cenozoic). On these islands and in the water around them were placed the life-sized models of 14 different extinct animals, from a Carboniferous amphibian to dinosaurs and a giant deer from the Ice Age.

Just before the Palace reopened, Hawkins arranged one of the biggest publicity coups in the history of science by holding a dinner party within the mould of the giant *Iguanodon* on New Year's Eve, 1853. Engraved illustrations of the dinner appeared in newspapers and journals around the world.

Somehow, a large number of these models have survived the intervening years and the destruction by fire of the Crystal Palace itself.

and arid conditions. Land-living reptiles spread out across the continents and began to diversify, with the previously dominant mammal-like reptiles now competing with two other rapidly developing groups of reptile – the archosaurs and rhynchosaurs. More reptiles moved back into the seas,

producing fish-eating groups such as the ichthyosaurs, while others took to the air for the first time during the Late Triassic era, leading to the evolution of the pterosaurs.

Of particular importance were the archosaurs, because they included ancestors of the crocodiles, dinosaurs and pterosaurs.

One was the Early Triassic *Proterosuchus*, a crocodile-like predator about 5 ft (1.5 m) long, whose short limbs and sprawling way of moving meant that it was not fast, but had enough speed to catch and feed on smaller reptiles. The smaller *Euparkeria*, which measured about 1½ ft (46 cm),

ANCIENT SMILE *Phytosaurs such as the* Nicrosaurus, *probably used the same tactics for hunting as modern fish-eating crocodiles.*

appeared slightly later but had a number of more advanced features. Its jaws were filled with the long teeth of a carnivore, firmly set in individual sockets for increased biting and holding power. Its long limbs and more flexible ankles may well have made it faster than other animals – whether it was travelling on two or four feet.

At the end of the Early Triassic era, around 250 million years ago, the archosaurs divided into two major groups: one led to the dinosaurs, pterosaurs and eventually the birds; while the other gave rise to the crocodilians and, ultimately, to the crocodiles we know today. Some of the early crocodilians, such as the 8 ft (2.4 m) long *Parasuchus*, from India, were primarily aquatic predators similar in many ways to living crocodiles. Others, such as the impressively sized 23 ft (7 m) *Saurosuchus*, which was alive in Argentina during the Late Triassic period, were long-legged, active terrestrial predators, capable of attacking almost any other animal of the time. Clearly the crocodilians had hit upon a very successful basic body plan and biology, which has allowed this ancient group to survive while the dinosaurs came and went.

LEADING UP TO THE DINOSAURS

Thecodont reptiles such as *Ornithosuchus*, which lived as far afield as Scotland and South America during the Late Triassic era,

preceded dinosaurs and were very similar to them.

Measuring up to 11½ ft (3.5 m) long, these animals had slender limbs, with longer legs than arms, and were probably capable of walking on either two or four feet.

Herrerasaurus, a near contemporary from Argentina dating from the Late Triassic, was of a similar size to *Ornithosuchus* but had the robust build of a fully bipedal predator. Its more advanced hip structure allowed it to stand upright on two legs, in a similar stance to that of bipedal mammals, and its ankle had developed into the strong, weight-bearing joint of a true dinosaur. The feet had elongated toes for running, with an enlargement of the three middle ones. The three-toed footprints they produced misled some scientists in the early 19th century into thinking that at some time in the past there had been a race of giant birds on Earth.

From an ecological point of view, it is curious that the first true dinosaurs were carnivores – it seems more logical that evolutionary changes in animals would start with the commonest plant-eating species at the

BONES OF EVIDENCE *The discovery in 1987 of this* Herrerasaurus *skull (right) in Argentina gave the first idea of its structural details. A predator around 220 million years ago, this thecodont reptile was able to stand upright on two legs (far right).*

bottom of the food chain. Yet the history of tetrapods tends to show the reverse – time and time again new groups start with carnivores. Perhaps this is less surprising in reptiles as the basic form of their teeth, relatively simple cones, are more easily adaptable for eating meat than plants. Many herbivorous dinosaurs never developed grinding-type teeth to help with predigestion. As they ate they swallowed stones, which lay in their stomachs and helped digestion by forming a kind of gastric mill.

DINOSAUR DIVERSITY IN A CHANGING WORLD

From a fairly minimalist start around 230 million years ago, with the earliest examples, such as *Herrerasaurus*, being very much in the minority (perhaps only 1 per cent of all animals), the dinosaurs developed within 20 million years to become the most important of the terrestrial vertebrates by the end of the Triassic era. In doing so, there is no doubt that they 'edged out' the mammal-like reptiles – but exactly how and why the takeover happened is still a matter of debate. Part of the argument involves the wider issue of whether the dinosaurs were warm-blooded and whether that conferred some advantage. The extent of that advantage depends on whether the mammal-like reptiles were also warm-blooded.

A further possibility is that there may have been at least one other extinction event at the

end of the Triassic period. As opportunists, the dinosaurs may have been able to take advantage of this in order to expand into the vacated niches. However, it may be that it was not a case of sudden replacement but gradual changes in a number of groups over tens of millions of years. Dinosaur characters did not all appear at once and it is perhaps our system of classification that forces or creates such problems. In nature, competition does not occur between different families of organisms but between different species and individuals.

During the Jurassic and Cretaceous periods, dinosaurs diversified into almost all land habitats, and they soon came in all sizes and forms. Herbivores ranged from small, fast-moving bipedal forms such as the 4 ft (1.2 m) *Thecodontosaurus*, to medium-sized forms that could walk on either two or four feet, such as the 23 ft (7 m) *Plateosaurus*, to massive, slow-moving and essentially quadrupedal animals such as the 88 ft (27 m) sauropod (semiaquatic dinosaur) *Diplodocus*.

Some dinosaurs were protected by body armour reminiscent of that worn by medieval knights, ranging from the bony plates of a 24 ft (7 m) *Stegosaurus* and neck shields of a *Protoceratops* (6 ft/1.8 m) to the horns of a 30 ft (9 m) *Triceratops*, the clubs of a *Pinacosaurus* (16½ ft/5 m) and the hideous spikes of a 13 ft (4 m) *Polacanthus*. All this weaponry was required to do battle with their enemies, the predatory carnivorous dinosaurs. These too ranged in size,

encompassing small, fast animals such as a 4½ ft (1.4 m) *Compsognathus* and a 9 ft (2.7 m) *Coelophysis*, right up to the slower 45 ft (13.7 m) *Tyrannosaurus*.

This diversification among the dinosaurs happened in what was initially a fairly stable global environment. The dry

climate of the Triassic era gradually gave way to a moister, warm climate stretching as far north and south as the Poles in the Jurassic era. Indeed, fossils of subtropical varieties of ferns and conifers have been found in rocks as far north as 60 degrees.

CLIMATIC CONDITIONS
The plants themselves went through a major change during Triassic and Jurassic times, as horsetails and club mosses left over from the Palaeozoic era were joined and eventually eclipsed by the true ferns and seed ferns, cycads and conifers.

By the Early Cretaceous times, these varieties were growing alongside the plants that would eventually come to dominate the plant life of the modern world: the flowering plants or angiosperms, as they are technically called. Like plants, the dinosaurs also flourished in these more temperate climates,

BODY ARMOUR *Thick, bony plates, spikes and tail clubs formed the effective defence strategy of the plant-eating ankylosaur,* Euoplocephalus.

and developed not only an enormous range of diversity but reached new heights – literally – in terms of gigantic size.

Four different groups of the large herbivorous and quadrupedal sauropods evolved from Late Triassic and Early Jurassic bipedal dinosaurs such as *Thecodontosaurus*, for example. These groups are classified on the basis of shared skull features. The camarasaurs and brachiosaurs both had an arch of bone between the nostrils, positioned quite high up and far back on the skull behind a well-defined snout. The diplodocids and titanosaurs, on the other hand, share long, broad snouts and a number of peculiar long cylindrical teeth at the front.

All four had large vertebrae to support their huge stomachs and intestines, but within a light, bony framework so as not to add any more weight to their considerable bulk. They travelled on pillar-like 'elephantine' limbs, which had to be stout

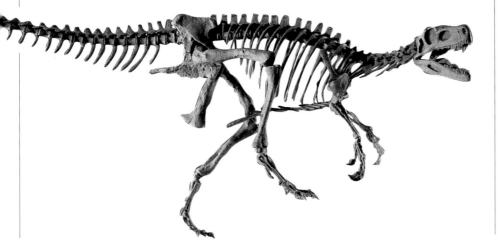

enough to prop the body up but at the same time articulated to be able to move it. The hands and feet were reduced to shortened weight-bearing structures.

LAND OF THE GIANTS

As dinosaur fossils have gradually been uncovered, it has become clear just how large some of them really were. Nevertheless, even scientists were in for quite a shock when the remains of the biggest of the sauropods – a plant-eating diplodocid called *Seismosaurus*, or the earth-shaker lizard – were found in the 1970s.

This plant-eating animal is estimated to have measured 128-170 ft (39-52 m) long and weighed about 100 tons, making it substantially longer than today's largest living creature, the blue whale, though not quite as heavy. A blue whale can weigh over 110 tons, but its bulk is supported by water.

RHINO DINO *Lethal horns and a bony neck guard were necessary protection for slow-moving ceratopsians such as* Triceratops.

TALL TAILS *Not tree trunks but the tail vertebrae of* Seismosaurus – *these fossil bones, discovered in New Mexico in 1979, led scientists to a giant among sauropods.*

What is so remarkable about *Seismosaurus* is that it was held up by its four limbs alone, linked by its pectoral and pelvic girdles and a backbone that stretched like an arch between them. Such a weight on these bones would have been close to the maximum stress possible for a land-living animal.

The first remarkable fossils of *Seismosaurus* to be found – a few tail vertebrae – were spotted in 1979 by some walkers near Albuquerque, in New Mexico. The vertebrae were so large that they looked more like a broken-up fossil tree lying on its side than bones, but one of the party, geologist Jan Cummings, was quick to recognise that these could be parts of an unusually large sauropod dinosaur. Although dinosaurs from the late Jurassic period had not been found in the sandstones of this region before, the curator of the New Mexico Museum of Natural History was persuaded to take a look. Once he was convinced, a team of scientists and a collection of hi-tech equipment were brought in to find more of 'Sam', the seismic sauropod.

Their first step was to establish in which direction any other fossil remains might be lying in the rocks below. Sophisticated techniques such as ground-penetrating radar were employed – and were partially successful in predicting where more bone material might be buried. The team also tried using high-resolution gamma-ray detectors to track down concentrations of uranium minerals, which appear to form in fossil bone.

Another mineral component of bone, hydroxyapatite, causes the fossils to appear fluorescent, so the scientists worked at night with ultraviolet light to see whether they could spot traces of bone. But it still required all the traditional excavating equipment of hammers, chisels, plaster jackets, pulleys and chains – along with hours of relentless hard labour – under the desert sun to recover what remained from the rock.

At the end of 15 years' work, they had managed to recover 37 vertebrae – a few from the neck and some from the chest with the ribs – the complete pelvis and part of the tail from the dirt and rock of the desert. A single block containing four of the smaller caudal – or tail – vertebrae weighed 1500 lb (680 kg) and the biggest block they excavated weighed 5 tons.

Unfortunately, the head and limbs of the giant beast were missing. The sauropod had been dead for quite a long time before it was eventually buried in the shifting sand of a river, and the missing parts must have been swept away by the current. However, an unexpected bonus was the discovery, in amongst the colossal bones, of clusters of well-rounded and polished pebbles – 240 altogether – ranging in size from 1 in (2.5 cm) to 4 in (10 cm) in diameter. These were almost certainly stomach stones – or gastroliths – which the animal swallowed to

continued on page 76

BIGGEST LEG UP *The scale of another giant dinosaur,* Ultrasaurus, *can be imagined from this full-size model of one of its front legs.*

HERD INSTINCT *Fossil trackways suggest that even large sauropods such as* Seismosaurus *moved about in groups, as elephants do today.*

LITTLE BIG HEADS *Relative to their body size, the skulls of giant sauropods are extremely small. Large eye sockets and peg teeth differentiate the skull of* Diplodocus *(left) from that of* Camarasaurus *(below left), with its huge nostrils in front of the eye sockets. Both were equipped for browsing in high branches (bottom left).*

help it digest its meals of plant material in the same way that some seed-eating birds, such as chickens, parrots and ostriches, still do today. These birds do not have teeth to help break up their food before swallowing it, but eat grit and stones instead. The particles become lodged in their gizzards, which are special chambers of their stomachs with relatively rigid and folded linings. Muscular contractions of this lining help break up the plant food material and the stones promote the process.

The idea that herbivorous dinosaurs without grinding teeth may have had a gizzard had been around for quite a while, but *Seismosaurus* was the first dinosaur discovery to feature fossil gastroliths that were so

closely associated with the area of the stomach. As is so often the case with fossil remains, however, the assumption that dinosaurs used stomach stones to aid digestion cannot actually be proved.

Given that the original populations of this dinosaur must have been in the order of thousands, and that the species lived for at least a million years, the total number of potential fossil skeletons is huge. Nonetheless, as yet, only one partial skeleton has been recovered by palaeontologists. This fragmentary preservation is typical of fossil records in general, but is especially true for those vertebrates, such as the dinosaurs, who lived and died on land.

MYSTERY OF THE LONG PEG TEETH

Detective work is essential in determining the lifestyles of ancient beasts from fossil evidence. For example, the fossil remains of *Diplodocus*, another giant plant-eating sauropod dinosaur, revealed that the animal had a peculiar, rake-like row of long peg teeth which showed unusual wear marks. Although at first the teeth were thought to have helped the animal tear up vast quantities of aquatic weed for food, now scientists think that these long-necked dinosaurs used their special teeth to rake and strip foliage from high branches. This idea is suggested by the unique patterns of

EATING UTENSILS *Rake-like cylindrical teeth (left) helped some sauropods strip foliage; while their thumb spikes (right) were used to grasp tree trunks for balance.*

wear seen on both the upper and lower sets of the round peg teeth at the front of the mouth. The teeth stuck out considerably from the jaw, with the flat, worn edges pointing outwards, so it is not feasible that the wear could have been made by tooth-to-tooth contact, even though the lower jaw could slide backwards and forwards.

These marks could have been caused by the animal browsing 50 ft (15 m) or so above the ground, among the high forest canopy. As it bit down on branches as close to a tree trunk as possible, and then pulled its head away, the peg teeth could have raked off all the foliage. When feeding from a high branch, the head pulled down and so the upper teeth became worn on the outside; when stripping a low one, the head pulled up and the lower teeth were worn on the outside. Such flexibility would have allowed the animal to gain as much food as possible from a single tree, a considerable advantage for a very large herbivore that would have had to spend most of its time eating.

Such an animal would have had to consume up to 400-600 lb (182-272 kg) of leaves a day to make up its energy requirements, especially with no cheek teeth to chew and prepare the food for digestion.

The peculiar thumb spikes of many sauropods at first were thought to be weapons. Now, it seems, some scientists believe they were more like grappling hooks, designed to help with high browsing. As the animal reared up on its hind legs to reach high into the forest canopy, the thumb spikes (up to 12 in/30 cm long) may have helped it to grasp the trunk of the tree and steady its 30-40 ton body as it fed.

Interestingly, the size and spacing of the hands of these large sauropods means that there would have had to have been Jurassic trees with diameters of up to 10 ft (3 m) if they were to have rested both hands on the trunk. Large tree fossils of this age are even rarer than dinosaur fossils, and as yet none of this size has been found.

THE MYTHOLOGY OF THE GIANT CARNOSAURS

In many ways, dinosaurs have replaced medieval dragons in popular mythology – as the word 'dinosaur' implies. It was coined by Sir Richard Owen in 1842 and is taken from the Greek *deinos*, meaning 'terrible', and *saurus*, meaning 'lizard'.

Of all of the dinosaurs, the ones that come closest to the Western image of the dragon wreaking havoc and breathing fire and brimstone were the giant, predatory meat-eating carnosaurs.

The giant carnosaurs must indeed have been grotesque animals, with enormous heads on short muscular necks, stumpy little arms and stiff, powerful tails balanced on top of long powerful legs. For top carnivores, these features would have provided the speed, agility and necessary equipment for them to kill their prey and consume it

THE BIG ONE *Experts clean up the skull of the largest and most complete* **Tyrannosaurus rex** *ever found – in South Dakota in 1990.*

as fast as possible. Kills that were not consumed quickly could too easily be snatched by numerous scavengers.

As yet, it is unclear what role the arms played, as they could not reach even as far as the animal's mouth, but its muscular tail was almost certainly used for balance. Although the skull was huge, in most species it featured large 'window' openings to reduce the amount of bone and so provide a light construction. At the same time, it retained enough strength for the attachment of powerful jaw muscles. Such features link them with the other theropods and the most primitive birds, such as *Archaeopteryx*.

Generally, the carnosaurs dated from the Late Cretaceous era and their territories extended from North America across into Central Asia. Among the largest were

Allosaurus, a 40 ft (12 m) Late Jurassic carnosaur found in North America, and the infamous gigantic tyrannosaurs.

TYRANNOSAURS ON THE RUN

The *Tyrannosaurus*, the biggest terrestrial meat-eater of all time, measured over 46 ft (14 m) long and stood over 18 ft (5.5 m) high. How *Tyrannosaurus rex* caught its prey is a subject still under discussion, as scientists try to establish whether or not it could run, and even whether the animal was a hunter at all – it may have been a scavenger. Such a lifestyle would not have required so much speed, and there would have been less risk of injury.

When scientists first discovered tyrannosaur fossils, reconstructed them as skeletons and then tried to flesh them out as real animals, they soon realised that tyrannosaurs were not only large but also that the proportions of their bodies, especially their relatively enormous heads and almost useless tiny arms, were quite unlike that of any living creature. What impressed scientists was the sheer size and weight – well over 13 000 lb (6000 kg) – of what is regarded by most as a cold-blooded reptile. The nearest equivalent is the slow-moving elephant, which stands about 10 ft (3 m) tall

LETHAL WEAPONS *Although not particularly large as dinosaurs go, at 10 ft (3 m) or so tall, Deinonychus was armed with wicked sickle-shaped finger claws and a huge toe claw – making it a formidable adversary. It is thought that these predators cooperated in hunting down their prey, inset right, much as lions do today.*

and weighs around 11 000 lb (5000 kg). An elephant travels at a maximum of 20 mph (32 km/h) even though it is a warm-blooded mammal.

The faster an animal runs, or the higher it jumps, the greater the force on its feet as it lands. With this in mind, the dimensions of the leg bones can be used to calculate bone strength, which is related to body weight, and sets an upper limit on the amount of force that could have been applied through running. Such figures help to calculate the speed of an animal, while the relationship between leg length and stride means that stride lengths from fossil footprints can provide an independent check on any calculated speed.

The most convincing tests on the bones of *T. rex* suggest that the animal could move at a maximum speed of around 19 mph (31 km/h) – not at all fast. What is more, it has been calculated that if an animal as tall and heavy as *T. rex* tripped and fell, its front arms would have been useless at breaking its fall; the resulting impact of the skull on the ground (a force of about six times its weight) would have been sufficient to smash it. Consequently, tyrannosaurs could not have afforded to run or risk falling over because they would have suffered serious damage.

However, it would seem likely that tyrannosaurs were risk-takers in much the same way as animals alive today. Giraffes, for example, gallop at speeds of up to 24 mph (38 km/h), even though they risk breaking a leg and, as a result, fatal injury.

TRACKING THE DINOSAURS

Palaeoichnology, or the study of fossil footprints, has helped enormously to provide information about social behaviour and predator/prey relationships in dinosaurs and other animals. Sets of prints or track-

ways are also helpful, although establishing whose foot made which track is often incredibly difficult, as there are rarely any bones that can be linked directly to the prints. In part, this is because the type of rock which holds a clear trace of a track tends to be different to the best kind of rock for preserving bone.

Although palaeoichnology dates back 150 years, 'dinotrekking' has become something of a growth industry recently – largely because of studies over the last two decades in the western interior of North America. The region has just the right mix of a semi-arid climate – which exposes lots of rocks – and geological history to turn up some spectacular finds.

Since tracks can rarely be linked to body fossils, the study of fossil footprints is full of pitfalls. For instance, it can be hard

BALANCING ACT *The massive head and body of the Tyrannosaurus rex was counterbalanced by its heavy, stiffened tail.*

to distinguish between amphibian and reptile prints in rock laid down during the Late Palaeozoic period, but the mere existence of these trace fossils provides further understanding of the early tetrapods and the environments in which they lived.

The difficulty of interpreting fossil tracks is well illustrated by some Permian footprints from the Grand Canyon area, first described in 1918. There were no skeletal remains with the tracks, which were named *Laoporus*, meaning stone tracks, but they consist of sets of parallel footprints of a small, four-footed animal approximately

Euoplocephalus   Triceratops   Apatosaurus   Ceratosaurus

THE RIGHT TRACK *It is rare that fossil footprints can be linked to specific animal remains, because of the type of rock in which they are preserved. But they can provide extra information – about the social behaviour of a species.*

WHERE DINOSAURS WALKED *In Late Jurassic times, five sauropods ambled side by side across wet sand (facing page). Millions of years later, their journey is one of many recorded in the valley of the Purgatoire river, Colorado.*

16 in (40cm) long. At first, it was thought that the creature must have been an extinct amphibian, but then geological interpretation of the sandstone concluded that they had been desert sand dunes – making this interpretation unlikely. Then, it was noticed that some of the tracks began and ended abruptly, and changed from a forward to a sideways motion.

Experiments with living animals showed that salamanders make such tracks when they are in shallow water, as do lizards walking over sloping sand-dune surfaces. Differences between some of the tetrapod tracks have been interpreted as resulting from the same animal moving at different speeds and probably made on dunes by mammal-like reptiles who were well adapted for such an environment.

Dinosaur tracks have been found in many parts of the western USA, but the best trackways are to be seen in the valley of the Purgatoire river in south-eastern Colorado. This spectacular site boasts more than 1300 tracks mapped out on a single surface. The tracks include the prints of sauropods and theropods of different sizes, and are accompanied by an abundance of other palaeontological information which suggests that the site was the shoreline of a slightly alkaline lake. Plants, algae, snails, clams, crustaceans and fish remains have all been found nearby.

The site also provides evidence of the sociability of sauropods. Five parallel trackways were made by young adult brontosaurs, who were walking together.

DINOSAUR TREASURES IN THE HILLS OF MONGOLIA

There have been exciting developments in dinosaur studies in the last decade of the 20th century, mainly thanks to the

UPS AND DOWNS *Fossil tracks established that the **Iguanodon** could walk on either two or four legs, resolving controversy about how it moved.*

astonishing discoveries made in the Late Cretaceous sedimentary rocks of the semi-arid desert wastes of Mongolia.

It all started back in the 1920s, when palaeontologist Roy Chapman Andrews first led an expedition to the Gobi desert. There, he discovered several clutches of eggs as well as fossils of babies, young adults and fully grown examples of the small, ceratopsian *Protoceratops* dinosaur. It was

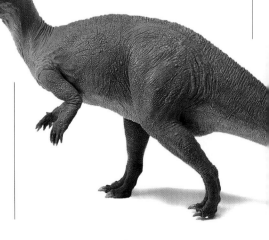

assumed that *Protoceratops* had laid the eggs, and this in itself attracted enormous public excitment. For the first time, dinosaurs could be seen as once-living animals that actually had babies and looked after them – just like any other 'real' creature.

Such spectacular finds also established Mongolia's Gobi desert as one of the great fossil-hunting grounds in the world. The success of the Andrews expedition to these Late Cretaceous sedimentary rocks (between 65 and 97 million years old) led on to joint Polish-Mongolian and Russian-Mongolian expeditions in the 1950s, 1960s and 1970s, and then to three American expeditions in the 1990s.

The most recent of these concentrated on the sandstone hills of Ukhaa Tolgod, in south-central Mongolia. These remote hills date back 80 million years, and are packed with the most outstanding fossils, few of

FOSSIL WONDERLAND *The Mesozoic riches buried in Mongolia's deserts (above) were first brought to world attention by the American palaeontologist Roy Chapman Andrews in the 1920s (right).*

which have been plundered. The finds uncovered so surprised the scientists on the expedition that they had the impression that the animals had given their last gasp just a few moments before their discovery.

The 1994 expedition to the region found one of the most important fossil treasure troves yet to be unearthed, and included 100 dinosaurs, 216 lizards and 187 mammals. More important still was the remarkable state of preservation of the skeletons, many of which were complete, with even the smallest, most delicate bones preserved. This is unusual, as the carcasses of animals are generally scavenged and scattered and then worn away by the processes of erosion and deposition before they are entombed in rock. Even then they can be damaged further during fossilisation and earth movements.

The explanation for this lies in the sediments in which the skeletons are interred. Scientists believe that the animals must have been overtaken by a sudden, catastrophic sandstorm to have been preserved so completely. Many would have died instantly, literally smothered by a blanket of sand that overwhelmed them before they had time to escape.

MAMMALS AND DINOSAURS COEXIST

An extra bonus of the Mongolian finds was that they provided palaeontologists with new and unexpected information about mammals of the Late Cretaceous era. This was a period when the mammals, although small in size, were diversifying and increasing in numbers and yet very little is known about them. No mammal skulls of this age have yet been found in the whole of North America. The skulls from the Gobi desert could help to establish how the different mammal groups related to one another and to the dominant dinosaurs – although it will be many years before the scientists finish their research on such a massive haul.

From the evidence uncovered, it seems as though placental mammals similar to those alive today lived alongside marsupials

MOMENT OF DEATH Buried and fossilised in the place where it died, this Mongolian Protoceratops is remarkably well preserved.

and an archaic extinct group called the multituberculates. The latter were by far the most abundant – 169 of the skulls found belonged to this group – and were probably plant-eaters. Another 18 skulls belonged to placentals and marsupials, which were insect-eaters. This sort of pattern is found in communities of small mammals today, where large numbers and kinds of plant-eating rodents dominate insect-eaters.

The discoveries raise similar questions about the different types of dinosaur around at this time. The first dinosaurs found in the Gobi were primarily plant-eaters, especially the famous *Protoceratops* 'family'. By contrast, the dinosaur finds of the 1990s are dominated by 39 skeletons belonging to at least seven

WRONGLY ACCUSED Even though it is now believed that Oviraptor *is innocent of egg stealing, it is stuck with its original scientific misnomer.*

different groups of carnivorous theropods, including eight oviraptors, several dromaeosaurs, a tyrannosaur-like skeleton and the dinosaur-type bird, *Monoykus.*

How all these different species lived and died together is as yet unknown, although further finds may unravel the mystery. In the 1920s, the discovery of a then unknown type of two-legged dinosaur in Mongolia near to a clutch of ceratopian eggs

NESTING INSTINCTS *Among fossil treasures uncovered in Mongolia were dinosaur eggs containing embryonic bones (above) and even a fossil nest with the arms of the parent Oviraptor still wrapped around its eggs (below). This was the first solid evidence of advanced reproductive behaviour in dinosaurs.*

led scientists to conclude that *Oviraptor philoceratops*, whose name means 'egg-thief with a love of ceratopian eggs', was an egg stealer. In the 1990s, tests have proven that the eggs were quite probably the *Oviraptor's* own, and other embryonic remains in the nest appear to be those of a third type of dinosaur. This hints at the possibility of cuckoo-like parasitic behaviour by dinosaurs, although this cannot be proved.

THE DINOSAUR NEST

Dinosaur eggs have been neglected, generally, since their first discovery in 1859, but now modern scientists are realising the great potential in fossil eggs as a source of biological information about how the dinosaurs once lived. This fascinating field of dinosaur research is still in its infancy, but it has established that nest sites and egg-laying patterns can reveal a lot about their social behaviour. Dinosaur eggs are being found in increasing numbers

all over the world – from Shabarakh Usù in Mongolia to Algorta in Peru and Cirencester in England. Whole eggs have been found, occasionally in patterned clusters at

THE FIRST EGG

Although scientists can surmise from the skeletal evidence available that egg-laying reptiles, or amniotes, evolved during the Late Carboniferous period, no fossil eggs of this age have been found yet. To date, the oldest known fossil reptile egg was found in Lower Permian terrestrial sediments of West Texas. It was laid roughly 280 million years ago by an unidentified reptile, and it is the first clear indication of true reptile egg-laying amniotes in the fossil record. Slightly younger fossil eggs found in rocks in southern Africa dating from the Permian era are thought to have been laid by dicynodont therapsids.

the original nest sites, as well as fragments of shell. A few extremely rare specimens have been found to contain the fossil remains of embryo dinosaurs. The eggs vary greatly in the details of their structure, which reflects the differences in reproductive biology from species to species.

Finds of embryos and nestlings have helped scientists to understand the growth rates and processes of the dinosaurs. One example of this is the first reliably identified sauropod embryo, or unhatched baby, of *Camarasaurus*, found in the Morrison Formation in Colorado. When fully grown, this plant-eating reptile stretched 60 ft (18 m) long. The baby measured about 3 ft (1 m), and was curled up in a fairly large egg with a capacity of about 12 pints (7 litres). The real significance of this find, however, was that it contradicted previous suggestions that the sauropods gave birth to live young.

In evolutionary terms, the development of the shelled or 'amniote' egg was one of the greatest advances in the history of life on Earth. Egg laying allowed the reptiles to become fully land living and totally independent of water for the first time. Now, both reptiles and birds lay eggs, not to mention that peculiar small group of egg-laying mammals, the monotremes, which includes the duck-billed platypus.

Do Giant Animals Lay Giant Eggs?

How big would a dinosaur egg need to be? A hen's egg has a volume of about 2 fl oz (57 ml), and that of an ostrich, about 3 pints (1.7 litres). If the proportions between the egg and its maker stayed the same, the egg of a 65½ ft (20 m) sauropod weighing 50 tons would have to be enormous, with a capacity of about 154 gallons (700 litres) and a weight of around 1540 lb (700 kg) – the equivalent of a small car. The shell would have to be incredibly thick to contain the weight, and the baby dinosaur would need a sledgehammer to break out.

Obviously, there is a fundamental error in these scaling-up calculations. The largest living reptile is the crocodile, and even a one-ton specimen lays smaller eggs than

BASEBALL-SIZED EGGS This reconstruction of a Maiasaura's nest and its hatchlings was based on finds from Montana, USA, made in 1978.

UNBORN BABY Inside a fossilised dinosaur egg can be seen the skull, lower jaw, hooked claws and vertebrae of a therizinosaur embryo.

those of an ostrich. To get a measure of the largest known viable shelled egg, it is worth looking at the birds. *Aepyornis*, the extinct elephant bird of Madagascar, laid an egg more than 11 in (28 cm) in diameter and 16 pints (9 litres) in volume.

Even this was too big for the dinosaurs, it seems, as nests, eggs and hatchlings found in Montana, North America, revealed that the eggs of *Maiasaura*, a 29 ft (9 m) hadrosaur, were no more than 4½ in (11.5 cm) long.

The biggest dinosaur eggs found so far are those of *Camarasaurus*, which were 9½ in (24 cm) in diameter – the size of a football – and held up to 12 pints (7 litres). An egg this size could have contained a tightly curled-up baby dinosaur measuring about 3 ft (1 m) and weighing perhaps 16½ lb (7.5 kg), the size of a smallish dog.

Camarasaurus was not a huge dinosaur – it was about 60 ft (18 m) long when fully grown – and as yet, the egg of a giant sauropod, for example, is yet to be found. Even

then it is unlikely to be much bigger than this. Basic scientific principles dictate that there is a maximum size for shelled eggs laid on land. The larger the egg, the thicker and heavier the shell has to be. As eggshells have to be thin enough both to breathe – to allow the diffusion of gases – and for the fully grown embryo to break out of, football-size eggs are perhaps the maximum practical size that can be achieved. If so, the largest dinosaurs would have had to care for extremely small babies in proportion to their full-grown size, similar to the care large alligators provide for their tiny hatchlings today.

Research continues, and will doubtless reveal more answers. In the meantime, new investigative techniques have enabled dinosaur and fossil turtle eggs from China to be opened without harming the 1-2 in (2.5-5 cm) skeletons of complete articulated and disarticulated embryos inside. Such tiny bones are so delicate that they are never well preserved in ancient rocks, but they do allow a detailed look at the early development of animals that disappeared so long ago.

MONSTERS OF THE DEEP

Long before the existence of dinosaurs, sea dragons were conducting their own reign of terror in the dark ocean depths. Cruising through the waters, ambushing anything in their paths, these beasts are now the stuff of legends.

Scotland's beloved Loch Ness Monster is a creature steeped in ancient mythological history, and the earliest discoveries of spectacular fossil remains were not those of dinosaurs at all but sea dragons or monsters of the deep as embodied in the popular imagery of 'Nessie'.

Such finds can be explained by the geology of the Earth at the time such marine animals were in existence. About 200 million years ago, a shallow but widespread subtropical sea flooded much of the area of what is northern Europe today. As the whole region was near the Equator at that time in history, the seas were subject to high evaporation rates. A continental shelf protected the shallow seas from strong tides or storms, so most of the sediment that was deposited on the seabed was fine-grained, calcareous – or chalky – and muddy. These conditions led to low oxygen levels in the sediment and prevented the development of much life on the bottom of the sea, with the exception of some shellfish that had managed to adapt. In the warm waters above, however, it was a different story, with an abundant variety of fish and free-swimming cephalopod ammonites and belemnites acting as food for larger vertebrates.

Creatures that died but were not eaten sank to the bottom of the sea and into the soft, sea-floor mud. Once there, they were slowly covered by more fine mud, which helped to preserve them, and, in time, these sediments gradually turned to stone. This process involved complex chemical changes within the sediments. The Jurassic sea-floor mud was transformed into alternating sequences of limestone and shale, which are now exposed in the coasts of Dorset and Yorkshire and inland hills throughout Europe.

For hundreds of years, these rocks were quarried for stone, which was then used to build the great Norman castles and medieval cathedrals. In the late 18th century, there was an increased demand for rock as a raw material for buildings, road stone and the manufacture of bricks and lime, and new sites were excavated

PICTURE THE PAST *A 19th-century reconstruction of a Jurassic tropical lagoon shows marine and flying reptiles and the bird Archaeopteryx.*

BEAST OF MAASTRICHT *The discovery, in 1786, of a giant reptile skull in a Dutch chalk quarry captured the interest of scientists, and artists.*

throughout France, Germany, Holland and England. These quarries exposed rocks that were predominantly Mesozoic and Tertiary in age, and were largely formed from marine sediment.

Landowners began to realise the potential value of the raw materials that lay beneath their green fields, and turned to the only people who could advise them of their worth: the small but growing band of professional civil engineers, surveyors and geologists. It was inevitable that this increase in geological activity would lead to significant finds of fossils.

Intellectuals of the time were stimulated by the evidence that creatures had lived in those ancient seas, particularly since they were discovered before Darwin had proposed his revolutionary ideas on evolution. In the late 18th-century, pre-Darwinian world, most people still generally believed in a Bible-based version of creation and the concept of extinction was unknown.

THE FIRST MONSTER 'SAURIAN'

The most spectacular of these early discoveries were the various 'saurians', as they were generally referred to at the time – their name was taken from the Greek for lizard. The idea of these extinct marine

reptiles stirred the curiosity of the late 18th-century natural philosophers (the word 'scientist' was not coined until the 19th century) when the first fossils surfaced in Maastricht in the Netherlands.

In 1766, some large fossil jaw bones were discovered deep within the Upper Cretaceous chalk rock of St Pieter's Mountain in Maastricht. Two decades later, a much better preserved skull more than 3 ft (1 m) long was found in the same place, and became Europe's scientific sensation of the 1780s – not to mention farther afield. At the time, the great American scholar and diplomat Benjamin Franklin was living in Paris, where he had been posted as the first ambassador to France from the newly independent America. His interest in natural philosophy makes it likely that he was aware of the find.

The ownership of the fossil had been contentious from the start, but went on to provoke a chain of events that included international piracy. The miner who discovered it sold it to a German military surgeon, but in turn he

TESTING THE WATER *Among the oldest marine reptiles were mesosars, which sieved plankton between numerous thin, interlocking teeth.*

was successfully sued by the owner of the land where it was found, Canon Godin. Once Godin had taken possession of the fossil, he displayed it in a glass case in his chateau, where it became an object of curiosity and pilgrimage for some 15 years for scholars and naturalists of the day.

IS IT A CROCODILE?

Among them was Dutch naturalist Pieter Camper, who published his analysis of the fossil in the 1786 Transactions of the Royal Society (a British institution, which was founded in 1645 and is the oldest continuing scientific society in the world). Camper pointed out features that differed from those of living crocodiles and concluded that the fossilised skull actually belonged to a whale. His conclusion provoked a long

running academic argument with scholars, but especially the French – who insisted that it was a crocodile.

In 1795, during the Napoleonic wars, Napoleon's Republican armies besieged Maastricht. But such was the fame of the 'grand animal of Maastricht', that the French general Pichegru ordered his gunners to spare the chateau. In fact, Godin had already hidden the specimen elsewhere to prevent it falling into French hands. When Pichegru put up a reward of 600 bottles of wine, the fossil soon surfaced, however, and was taken to Paris. It is still there today, in the Muséum National d'Histoire Naturelle.

French scholars were still insisting that it was a crocodile in 1799, but only a year later Pieter Camper's son showed, from a detailed study of its bone structure, that it was neither a whale nor a crocodile but a giant lizard. He wrote to the great French comparative anatomist and palaeontologist, Baron Georges Cuvier, telling him of his conclusions. Cuvier subsequently wrote several accounts of its structure supporting Camper's diagnosis, but such was Cuvier's fame that Camper's role in producing the right zoological diagnosis was soon forgotten. Cuvier is generally credited with the

HOSTAGE OF WAR *The 'grand animal of Maastricht' – Mosasaurus hoffmani – was captured in 1795 and has been held in Paris ever since.*

correct identification, although it was an Englishman, William Conybeare, who named the animal *Mosasaurus* in 1822.

Very little further work was done on the original specimen until the late 20th century, when analysis using modern techniques drew comparisons with other mosasaurs – the largest of all predatory marine reptiles. These awesome animals grew to over 55 ft (17 m), had jaws up to 5 ft (1.5 m) long, and spread with enormous success throughout the oceans of the Upper Cretaceous world. During this time, 20 different genera and 70 species developed worldwide.

TERROR ON THE CRETACEOUS SEAS

One of the last and most advanced of these species was *Mosasaurus hoffmani* whose huge size meant that most other marine animals were potential prey. Its teeth had the most

advanced cutting edges of any marine reptile, and each tooth crown had numerous cutting or breaking facets which were capable of both crushing and cutting. The jaw had become mechanically more effective and its muscles more powerful than those of its ancestors, and the skull was more rigid and stronger. It is believed that the mosasaur's sense of smell was not particularly good but its eyes were quite large, indicating generally good sight (although limited binocular vision meant that depth perception would have been poor).

This enormous reptile's speed was generated by sideways undulations of its long, dragon-like tail, while two pairs of paddle-like limbs steered the massive body where it needed to go. Their shape indicates that the creature turned by altering

UNDERWATER KILLER *Large teeth and eyes, supported by a ring of bones, characterised Prognathodon solvayi, a deep-water mosasaur.*

the angle of the appropriate paddle, which would have put a huge amount of force on the articulating joint.

All of these structural and anatomical features indicate that the animals were fast-swimming hunters in surface waters, but there is also independent evidence of their savagery. Teeth marks found on the fossilised shell of a giant turtle, *Allopleuron hoffmani*, and healed jawbone fractures in a number of mosasaur specimens, show that they were indiscriminate predators. They even engaged in potentially lethal male-to-male combat with their own species. The fact that some of them survived a fractured jaw shows that they must have had rapid recuperation rates, rather like modern-day alligators and crocodiles.

The disappearance of the mosasaurs at the end of the Cretaceous period is part of the enigma of the 'K-T' extinction event. (Responsible for wiping out all dinosaurs, in 'K-T' the 'K' stands for 'Kreide' – the

German word for chalk thus Cretaceous – and the 'T' represents 'Tertiary', the period that followed afterwards.) As a group, the mosasaurs had an extraordinarily brief existence of only 25 million years, and were just at the point of an enormous expansion when their extinction occurred about 65 million years ago. The hole left in the world's oceans was soon filled by marine mammals in Tertiary times.

THE WONDERS OF LYME REGIS

Despite the initial interest in, and notoriety of, the Maastricht mosasaur, the scientific focus soon changed to where the new

NO STONE UNTURNED *The Jurassic cliffs around Lyme Regis in southern England have been famous fossil hunting grounds for 200 years. The complete bony fish* **Pholiodophorus** bechei *(left) is just one of many remarkable finds made in the fossil-rich limestone and mudrocks.*

discoveries were being made. The best of the new 'monsters of the deep' were being found in the Lower Jurassic rocks around Lyme Regis, in the south of England.

Fossils were, and still are, abundant in the Liassic limestones and shales that form the local cliffs in the region. An extraordinary variety of fossil clams, coiled ammonites, some fish and rare vertebrates, were revealed by rock falls as storm waves constantly undermined the cliff faces, causing them to destabilise and collapse. Tons of rock strata containing fossils fell down onto the beaches, as they do today, where *continued on page 92*

Marine Reptiles

The seas of the Mesozoic era were inhabited by a variety of large and often ferocious reptile predators. The evolution of these monsters of the deep is still very obscure, but they all must have arisen from earlier terrestrial reptile groups. The most ancient, the marine crocodiles, were also the only group to survive the Late Cretaceous extinctions.

The dolphin-like ichthyosaurs appeared somewhat mysteriously at the beginning of the Triassic period, some 250 million years ago, without any clear ancestors. They survived for over 170 million years, until the Late Cretaceous era, and then disappeared abruptly along with other marine reptiles, such as plesiosaurs and mosasaurs, and the pteranodontid flying reptiles. This extinction event occurred 8 million years before the dinosaurs were wiped out. Plesiosaurs evolved slightly later, at the end of the Triassic, from the solely Triassic nothosaurs – another group of marine reptiles. Three or four main groups developed, with a considerable variety in body size – anywhere from 6½-46 ft (2-14 m) long – and proportions. These differences, from long necks and small skulls to short necks and massive skulls, were related to different modes of life and feeding strategies.

The mosasaurs represented another group of large marine reptiles. They evolved from a quite different stock of reptiles, the lizards, in Late Cretaceous times. Ranging in size from about 3-55 ft (1-17 m), they were the most spectacular of the lizards and were the equivalent of today's killer whales, with heavy skulls and sharp teeth. The 20 or so different groups showed no sign of decline before they abruptly disappeared from Earth.

Marine crocodiles form an interesting and diverse group which includes other crocodilians, such as the mesosuchians (which were adapted for water dwelling). They included forms such as the 6½-9 ft (2-2.7 m) *Geosaurus*, whose remarkable aquatic adaptations paralleled those of the ichthyosaurs. Its limbs had evolved into flippers, and its tail bent down into the lower

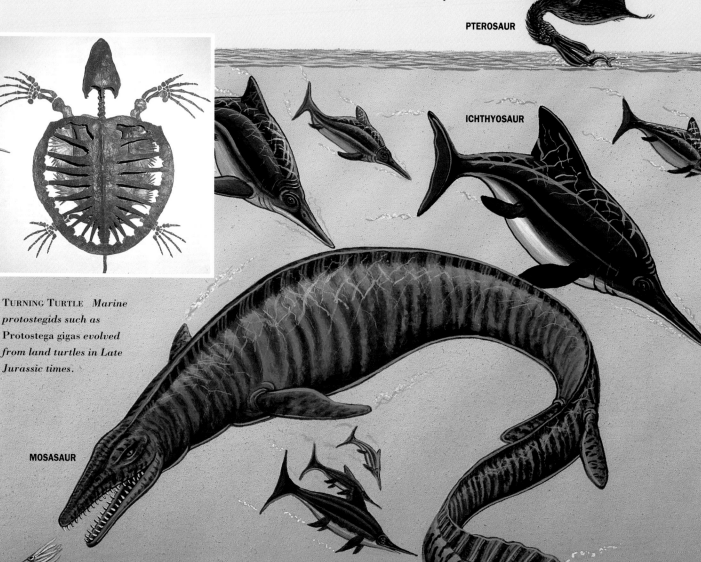

PTEROSAUR

ICHTHYOSAUR

MOSASAUR

TURNING TURTLE *Marine protostegids such as Protostega gigas evolved from land turtles in Late Jurassic times.*

lobe of a vertical tail fin and was balanced by an upper lobe. As a group, the crocodiles are still thriving.

Turtles have a highly modified reptile skeleton fused inside a two-part box shell, which itself is an integral part of the skeleton. The shell is made of bone, derived from enlarged ribs, and covered with horn. The group first evolved in the Late Triassic era and were mainly land-living until the Late Jurassic era (the period from which the first few specimens found in coastal sediments date). They probably fed on shellfish there.

The first fully marine turtles, such as *Archelon*, have been found mostly in North America and are Late Cretaceous in age. *Archelon* was a gigantic turtle which grew to 13 ft (4 m) and had a hooked, toothless jaw. The bone of its shell was lighter than that of earlier turtles, probably as an energy saving device, and its broad, paddle-like limbs produced a swimming motion that was a form of underwater flying.

This initial group of marine turtles disappeared at the end of the Cretaceous era but was replaced by new marine groups. Today's western Pacific loggerhead turtles are just one of several groups of large ocean-going turtles that return to land only to lay their eggs. Most of the rest of their lives is spent being carried by ocean currents across to southern California and then swimming back, against the current, to breed and nest in Japan and Australia.

The largest living turtles are the leatherbacks, which grow to 6½ ft (2 m) or more and can reach more than 1000 lb (454 kg) in weight.

DANGEROUS WATERS *Long before dinosaurs were discovered, the first real 'dragons' that captured popular imagination were fossil monsters of the Mesozoic seas. Many were discovered in Jurassic rocks exposed around the British coast near Whitby in Yorkshire and Lyme Regis in Dorset. These carnivorous marine reptiles included a variety of dolphin-like ichthyosaurs and long and short-necked plesiosaurs. The latter have provided the model for contemporary dragons such us Scotland's Loch Ness Monster.*

PLESIOSAUR

MESOSAUR

PLIOSAUR

SELLING SEASHELLS *More than 150 years ago, Mary Anning discovered some of the first Jurassic marine reptile fossils known. In this letter to geologist Dean Adam Sedgwick (above) she boasts that the scientist Reverend Conybeare is 'quite in raptures' with some of her finds.*

they were found by collectors or destroyed by the waves as they tumbled the slabs together in the rise and fall of the tides.

Many of the earliest of the magnificent fossils found around Lyme Regis were recovered by an uneducated Dorset woman, Mary Anning, and her family.

The peasants of 19th-century rural England were frequently impoverished and scraped a living as best they could. For the Annings, life became hard after the death of the head of the family, Richard Anning, in 1811. They were supported by the meagre benevolence of 'poor relief' and from what they could earn from collecting and selling fossils and curios to the tourists.

The tiny fishing village of Lyme had long had its share of summer visitors. Most were gentry who came to admire the seascape, but in the early 19th century the educated middle class were becoming interested in exploring the natural world.

THE INFLUENCE OF MARY ANNING

Typical of their time were the three Philpot sisters, avid samplers and collectors of natural objects, who first visited Lyme in 1806 and befriended the seven-year-old Mary. Over the following years they bought numerous fossil specimens from Mary and her family as, even at such a young age, she knew where to look and what to look for in the way of potentially valuable fossils.

In 1811, her older brother Joseph found the head of a 'crocodile', and the following year he dug the rest of the fossilised skeleton out of the cliffs – possibly with the help of Mary. The specimen was in fact one of the first ichthyosaurs to be found, and attracted a great deal of attention when it was sold to the Lord of the Manor, Henry Henley, for £23 (a great deal of money in those days). It is now to be found in the Natural History Museum in London.

More and better finds of 'saurians' were made, including one particularly fine ichthyosaur which was sold in 1819 for £100 (about £20 000 today), and the fame of the Annings spread among scholars and collectors. Although Mary probably had little or no schooling, she learned to read and write so that she could communicate news of finds to those who might be interested –

Lyme was a long way from London by horseback or carriage, and off the beaten track.

Most of those she wrote to were well-educated gentlemen, and the stream of visitors to her small shop included eminent geologist Henry De la Beche, William Buckland (first professor of geology in Oxford), Louis Agassiz (a famous Swiss geologist) and even the King of Saxony. There, they could see her latest specimens and discuss them in scientific detail. Many remarked upon her knowledge of anatomy, especially that of the marine reptiles. She was not afraid to engage the likes of Professor Buckland in a dispute over fossil interpretation.

One of her best finds was that of a complete 9½ ft (2.8 m) plesiosaur in 1823. Earlier, Henry De la Beche and the Reverend William Conybeare had realised that bones thought to belong to an ichthyosaur were in fact those of a quite different reptile. Although they called this creature *Plesiosaurus* (Greek for 'near lizard'), other scholars doubted its

LAYERS OF PREHISTORY

The small town of Lyme Regis sits in a hollow above a wide, shallow bay. The bay opens on to the English Channel but is protected from stormy seas by the Cobb, a long, curving sea wall. The town is confined to the only stable ground in the area and is surrounded by cliffs that are notorious for their landslips. These slides and slumps result from rainwater weakening then lubricating the alternating layers of limestone and shale. The Lower Liassic cliffs then collapse seawards.

These cliffs are internationally famous for the richness and the diversity of their fossils, with their layers of limestone and shale subdivided into 20 successive ammonite zones – each of which is about 23 ft (7 m) thick. Each layer took a million years to build up.

Now the cliffs are officially protected as a Site of Special Scientific Interest, and the local Lyme Regis Philpot Museum – named after the Philpot sisters – displays the local geology and the history of the Anning family.

LAND MARK *These strata of clay and lime, which form the fossil-rich cliffs at Lyme Regis in Dorset, were originally deposited on the seabed around 200 million years ago.*

existence until Mary proved them right by finding this complete specimen. The following year, 1824, Conybeare wrote an account of the new fossil and concluded correctly that it was a marine reptile that swam slowly with its flippers, rather like a turtle. He considered that its long, flexible neck compensated for its small head and weak jaws by being able to bend quickly and snap up prey. Sadly, Mary Anning's social status and sex prevented her from entering

SEA DRAGON *This beautifully preserved* Plesiosaurus dolichodeirus *was found by Mary Anning in 1823.*

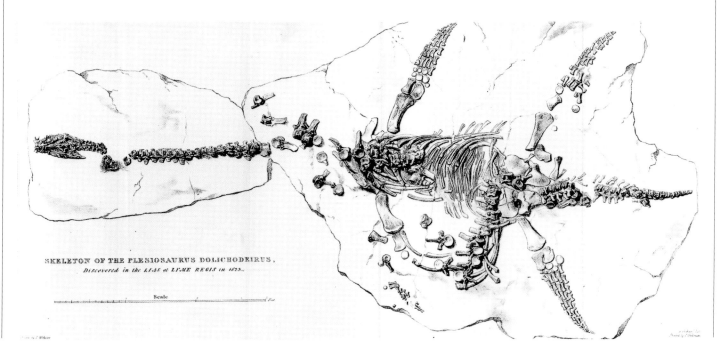

SKELETON OF THE PLESIOSAURUS DOLICHODEIRUS,
Discovered in the LIAS at LYME REGIS in 1823.

Scale

JURASSIC FISH

The warm shallow seas of the Jurassic era swarmed with a great diversity of fish. Most resembled the fish of today generally, but they were essentially different in detail.

Most modern fish groups belong to the true bony fish, teleosts, which are characterised by very thin body scales. In Jurassic times, the teleosts had only just begun to evolve and were overshadowed by an older group of fish, the holosteans, which had larger, heavier bony scales. Most were less than 3 ft (1 m) in length and, although many of them became extinct during the Cretaceous period, some species, such as the pike, survived until today.

Cartilagenous sharks and rays were present in the seas and gradually increasing in numbers, and a few surviving lungfish and coelacanths were also around. Another important ancient group were the chondrosteans, from which the sturgeons of today descend. Some of these grew to a

CRUSTACEAN CRUNCHER *The teeth of* Dapedium – *a common Jurassic bony fish from Lyme Regis, southern England – are adapted for feeding on crustaceans.*

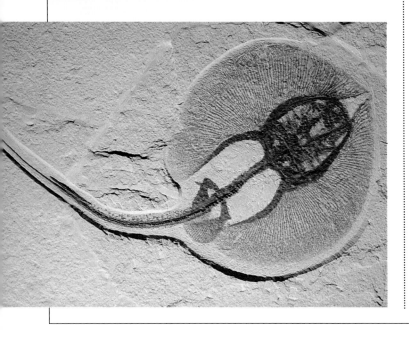

WATER WORLD *The Upper Jurassic tropical seas teemed with fish such as this* Gryouchus, *found in Solnhofen, Bavaria.*

considerable size: the magnificent *Gyrosteus mirabilis*, originally found in the fossil-rich Upper Liassic rocks of Whitby, on the north Yorkshire coast, is estimated to have grown to a length of over 16 ft (5 m).

Lyme Regis is one of the best fossil sites in the world for Lower Jurassic fish – during the 19th century the fossils of around 50 new species were found there. Many of these were well-preserved entire specimens, and they can still be seen in museums around Britain today.

STING IN THE TAIL *Rays with tail barbs first appeared in the Jurassic era but became common as fossils in the Tertiary era, as in this* Heliobatis *specimen.*

Squaloraja (1828) and many other invertebrate shells. She was also possibly the first person to correctly identify the phosphatised fossil fish and reptile faeces that are fairly common in the Liassic shales.

Very few of her fossils in museum collections acknowledge her as the finder, and it is only now, when museum curators are more interested in the history and provenance of their specimens, that her discoveries are being identified. Only one of the five most important British institutions to purchase her specimens – Oxford University Museum – has a direct record of a specimen originating from her. It was customary at the time to record the name of the donor only. Oxford University Museum also has the Philpot sisters' magnificent collection, while Cambridge University's Sedgwick Museum has several of her prize ichthyosaur fossil specimens.

Her income fell as she grew older, thanks to a general drop in interest in fossils, and breast cancer forced her to give up the hard life of 'fossicking' out on the beach. Fortunately, the scientific community had not entirely forgotten how much they owed her and, at the meeting of the British Association for the Advancement of Science in 1835, £200 was raised by private subscription. Buckland persuaded the prime minister, Lord Melbourne, to add a further £300 in 1838 and together, these sums bought Mary Anning an annuity of £25. She died in 1847 at the age of 48.

There is an old English tongue-twister: 'She sells seashells on the seashore. The shells she sells are seashells, I'm sure. For if she sells seashells on the seashore, then I'm sure she sells seashore shells.' It may well be

the developing world of 19th-century science, which was dominated by university-educated, middle-class men and which was rapidly becoming professionalised.

How much of her observation found its way, unacknowledged, into the scientific books and papers that Buckland and others wrote is a matter of argument, but she felt that her discoveries had been used without her role being recognised. What is undisputed is that she found at least three complete ichthyosaurs (1818, 1821 and 1830); two plesiosaurs (1823 and 1830); the cephalopod *Belemnosepia*, with its fossilised ink sac preserved; the first British pterodactyl (1828); the cartilaginous fossil fish

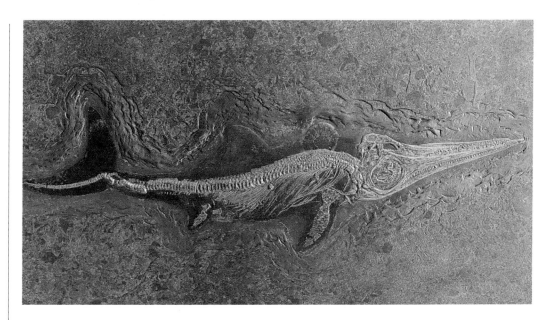

WHALE OF A TAIL *The fossilised body outline of the ichthyosaur* Stenopterygius *shows that the tail is bent down naturally.*

and just such a ring is present within the large eye socket of this early fossil.

At first, scientists thought that the ichthyosaurs had long, sinuous, serpentine tails – their name means 'fish-reptiles' – and that their seal-like paddles were used for hauling themselves ashore. As reptiles, it was thought, the ichthyosaurs would have had to go ashore to lay their eggs. It was only when a specimen was found with juveniles within the body cavity that scientists realised their mistake. As with mammals, the fertilised eggs were retained within the body until the embryos were sufficiently developed to be born directly into the sea and fend for themselves. It is now generally accepted that the ichthyosaurs were ovoviviparous – capable of bearing live young from eggs

THE CONUNDRUM OF THE ICHTHYOSAUR'S TAIL

The ichthyosaurs were fast swimmers, propelled forward mostly by their vertical tail fins. With this sort of propulsion, the direction of the forward movement depends on

that this rhyme refers to Mary Anning, a testament to her knowledge of the cliffs at Lyme Regis and their fossils.

THE EMERGENCE OF ICHTHYOSAURS

Many marine ichthyosaurs of the Mesozoic era were like dolphins, adapted for rapid underwater cruising and the pursuit of small, fast fish. Most of these reptiles were around 9¹/₂-13 ft (3-4 m) long, although some

SEA BIRTH *Marine reptiles such as the ichthyosaur* Stenopterygius *could not go ashore to lay eggs. Instead they bore live young at sea.*

Triassic forms reached up to 49 ft (15 m). The oldest surviving substantial ichthyosaur – collected by Joseph Anning from Lyme Regis – has a skull over 3 ft (1 m) long, and bird-like, beaked jaws set with a fearsome array of crocodile-like teeth. The small, sharp cones would have been ideal for grabbing fish and cephalopods before swallowing them whole – as the fossilised stomach contents of some ichthyosaurs confirm.

Good vision and distance judgment are critical for active hunters and, in deep dark waters, this requires large, light-gathering eyes. A ring of bones helps to keep the enlarged eyeball within its socket and strengthens the eye during deep diving,

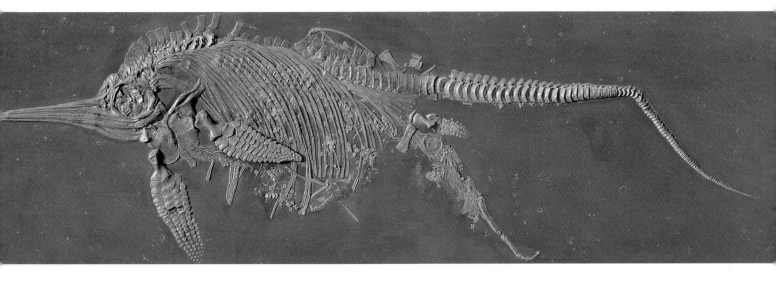

the shape of the fins and their relative flexibility. Traditionally, the ichthyosaur tail fin was interpreted as having worked in much the same way as a shark's, but in reverse. Whereas the shark's backbone bends up into the upper lobe, or section, of the two-pronged tail, and stiffens it, the backbone in ichthyosaurs can be seen to bend down into the lower lobe of the tail with the same effect.

When the first specimens of ichthyosaurs were found with the tail vertebrae bent downwards, it was thought that this must have occurred during preservation, as no living animal was known to have this backbone shape. In some specimens, the tail was even straightened out to make them look more authentic. It was not until some complete ichthyosaur skeletons were discovered in the Holzmaden quarries in southern Germany in the late 19th century that the truth was realised. These German specimens seem to show traces of a preserved body outline and have the peculiar broken-tailed appearance of so many of the fossils, with the backbone descending into the lower lobe of the tail.

As the preservation of these fossils was so good, it became clear that the bent backbone was an original and genuine characteristic of the ichthyosaur. The same

To the Bone   The skeleton of the whale-sized plesiosaur, Microcleidus, was extracted in its entirety from Early Jurassic rocks in Germany.

specimens also showed the presence of a large upper lobe to the tail fin, giving the tail a crescent shape which is reminiscent of that of a swordfish or tunny – both lobes being long, narrow and pointed.

It has now been shown that while there is a superficial similarity between the tails of sharks and ichthyosaurs, it is more accurate to compare the ichthyosaur's tail to that of some whales or other cetaceans (such as dolphins), and possibly to certain true fish such as the swordfish and tunny. Similarly, the swimming techniques of these ancient air-breathing reptiles is likely to have been closer to that of air-breathing whales and related cetaceans than to that of sharks.

Unlike sharks, marine creatures such as air-breathing whales have positive buoyancy – meaning that their bodies are lighter than water – at least while close to the surface. Ichthyosaurs, like cetaceans, had low density bones and this means that they were probably also positively buoyant. Diving would have been initiated by a downward flexing of the body, with increasing pressure reducing lung volume and buoyancy. The large, paddle-shaped pectoral fins would then have been quite adequate for adjusting swimming levels.

In fact, the ichthyosaurs might have been even more efficient swimmers than is at present realised. They were certainly among the earliest of the large marine vertebrates to achieve such a sophisticated, streamlined design.

PLESIOSAURS – A BLUEPRINT FOR THE LOCH NESS MONSTER?

Plesiosaurs were the most extraordinary marine creatures and do not really have a living counterpart – although reconstructions of Scotland's mythical Loch Ness Monster are often based on their form. They ranged in size from 6^1/$_2$- 45 ft (2-13.7 m) long and were well adapted for swimming, with four large, paddle-shaped limbs and a

Underwater Flight   Seen from below, the powerful paddles of the plesiosaur Rhomaleosaurus flapped it through the water like a bird.

long, flexible neck. Exactly how plesiosaurs swam has been a matter of speculation for some time. One theory suggests that they used a curious and powerful flying-like movement through the water, similar to that employed by turtles and penguins today. The flat paddle might have worked in much the same way as a bird's wing. Lift and forward thrust was generated by pushing down and backwards with a slight tilt from the horizontal, and then the paddle was tipped the other way on the up-stroke to generate an additional but smaller forward lift. During a complete beat, the tip of a paddle would have passed through a figure-of-eight pattern, producing forward movement at each stage of the cycle. However, analysis of some skeletons suggests that the construction of the hip and shoulder girdles may not have allowed such freedom of movement for the paddles.

SWIMMING LIKE A BIRD

A modified version of the plesiosaur's swimming technique suggests that the paddle tip followed a more restricted, crescent-shaped path — down and back and up on the main thrust, then forward and up on the return. Certainly, the biggest of the plesiosaurs must have been an impressive sight, 'flying' slowly and gracefully through the water. Most had long, flexible necks and relatively small skulls with long snouts. Like today's whales and dolphins, these marine reptiles had to surface every now and again to breathe, and they had a nostril set back from the tip of their snouts. In view of their size and slow swimming speed, they may well have been ambush hunters, using their long necks to snap up fish from passing shoals. Their jaws were full of closely interlocking crocodile-like teeth, which were well adapted for holding slippery fish in the mouth.

Short-necked plesiosaurs have come to be known as a separate group, pliosaurs, although they too could grow as big as 39 ft (12 m). Their relatively large, and sometimes massive, skulls were armed with a ferocious array of long, sharp teeth that often stuck out from their jaws. These pliosaurs were faster predators than most plesiosaurs, and probably hunted quite large prey – such as ichthyosaurs and other smaller plesiosaurs, as well as fish.

IN THE SWIM *Despite their considerable body size, plesiosaurs were mainly fish-eaters and had relatively small heads.*

THE RACE FOR THE SKIES

Gliding, soaring, diving, swooping – the Late Triassic

heavens teemed with fearful, sinister reptiles, some as big as

Second World War fighter planes. These unlikely aviators

would be ancestors to the more familar feathered varieties.

Insects first took to the air more than 300 million years ago, in Carboniferous times. Wherever plants grew in abundance, in forests and swamps, insects ruled the airways, unthreatened by any other flying creatures. But it was not long, in evolutionary terms, before the insects of the air were joined by the earliest flying vertebrates in the form of pterosaurs such as *Ptederodactylus*, who were airborne by the Late Triassic era, 220 million years ago.

By the succeeding Jurassic times, large self-propelled, heavier-than-air 'flying machines' had control of the skies for the first time in the history of life.

THE FIRST VERTEBRATE FLIERS

These flying machines were in fact vertebrate animals, but they bore no relation to the birds and bats that rule the skies today. These pioneer aviators were fish-eating

TAKING FLIGHT *The first known pterosaurs,* Eudimorphodon *and* Peteinosaurus, *were in flight around 220 million years ago.*

predators, reptiles who came to dominate not only the forests but also the coasts and surrounding seas. Collectively known as pterosaurs, they were to rule the air for 150 million years. They coexisted with their distant relatives, the dinosaurs, and perished with them at the end of the Cretaceous era.

Pterosaurs had short bodies, long necks, large heads, pointed jaws and, most importantly, hands with a strange elongated fourth finger. The long, tubular bones of this finger formed the leading edge of the wing and supported

a wing membrane, which stretched behind the hands and attached to the thighs or ankles. The wing membrane was composed of skin, reinforced and strengthened against damage by parallel, stiff elastic fibres. These allowed the wing to stretch for flight then shrink back to a manageable size when folded up.

From their early beginnings, the pterosaurs diversified to an extraordinary extent. They ranged in size from the

FLIGHT PLAN *Eudimorphodon (left) showed well-developed flight features such as a wing-supporting, elongated fourth finger. The hand (above) retained three grasping fingers with sharp claws.*

Eudimorphodon, which was about the size of a pigeon, to the giant *Pteranodon*, which had an elongated, crested skull of about 5¾ ft (1.7 m) and a wingspan of 26 ft (8 m).

Even the earliest of the pterosaurs, such as *Eudimorphodon,* appears to have been already well adapted for flight. It seems, from the fossil record, to have come from nowhere, without ancestors or transitional forms. As with the sudden appearance of marine reptiles, however, this is more likely to do with missing links in the rock record than a real occurrence.

To convert even a small and light reptile such as *Eudimorphodon* – with a skull that was only 3½ in (9 cm) long – into an effective flier meant adapting many of the features found in earlier reptiles. A special small bone in front of the wrist supported a forward extension of the wing onto the upper arm, for example, while the bones of the long tail were stiffened with special tendons, perhaps so that it could act as a sort of rudder during flight. Similar structures are found in dinosaurs, where a semirigid tail acted as a balance and sometimes doubled up as a weapon.

The variation in the form of the pterosaur skull was produced by different feeding habits. *Dimorphodon* had a short skull with small, sharply pointed teeth that may have been ideal for eating insects. By contrast, the long and rather widely spaced teeth of *Rhamphorhynchus* were probably used for holding fish caught on the wing while trawling the lower jaw in the water. The 500 long, comb-like teeth found in each jaw of the bizarre *Pterodaustro* must have been used for sieving microscopic organisms from the water.

Conversely, the *Pteranodon* of the Late Cretaceous period had no teeth at all. Like *Rhamphorhynchus*, it is probable that it hunted on the wing, and the fish that it caught were swallowed so rapidly that there was no need for teeth.

The *Pteranodon* ranged considerably in size, from those with wingspans of about 6 ft (1.8 m) to giants stretching to a fantastic

FINE-TOOTH COMB
Pterodaustro fed by sieving aquatic microorganisms with the hundreds of long, flexible teeth in its jaws.

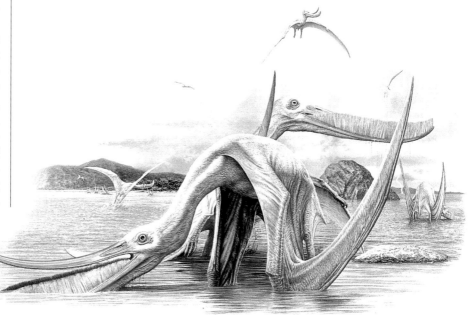

BIG BIRD *The pterosaur* Quetzalcoatlus, *with a wingspan of more than 36 ft (11 m), is the largest animal to have taken to the skies.*

30 ft (9 m) across, but even these were not the largest creatures ever to take to the skies.

GIANTS OF THE AIR

In the Late Cretaceous era, the largest of all pterosaurs evolved. The first evidence of the azhdarchids, as this group of outstanding animals is called, appeared with the discovery of *Quetzalcoatlus*, found in Texas in 1971. At first, the experts found its size hard to believe – its enormous wingspan was estimated at about 36 ft (11 m). This made *Quetzalcoatlus* by far the largest flying animal of all time, three times the size of the biggest bird alive now and closer in stature to Second World War fighter planes.

Now, it seems, there is evidence to show that a bigger type of pterosaur existed than even the monster *Quetzalcoatlus*. In the 1950s, a fossilised bone that had been

THE PRICE OF A PTEROSAUR

The value of well preserved vertebrate fossils has always been considerable, even since the days of Mary Anning. Today, public museums and private collectors vie with one another to acquire the finest specimens, either from auctions or specialist dealers. In 1995, the National Museums of Scotland paid £55 319 for a specimen of the *Rhamphorhynchus gemmingi* – rated by one expert as one of the best of its kind in the world. Originally the specimen was found in the 153-million-year-old limestone rock in the Solnhofen area of Germany.

found in Jordan in 1943 was taken to Paris where it was analysed by a French palaeontologist, Camille Arambourg. At the time, it was concluded that this was the wing bone of a pterosaur, but the fossil mysteriously

disappeared and so was not available for modern analysis. Recently, however, examination of a plaster cast of the bone made it clear that it was not a wing bone but a single neck vertebra. This has now been confirmed by expert palaeontologists, Dr David Martill and Dr Dino Frey, who in 1995 managed to track the fossil down to a storeroom in Jordan. From their work on the fossil, the geology team at the University of Portsmouth were able to estimate the total wingspan.

Some 65 million years old, the 24 in (61 cm) bone is laterally compressed and incomplete, but its full length is estimated to be about 30 in (76 cm). In all, the creature's neck would have been about 6½ ft (2 m) long – significantly greater than that of any other known pterosaur. It was found in sedimentary rocks of the same age as those which produced *Quetzalcoatlus*, but its size suggests that the original owner was in fact a Late Cretaceous pterosaur called *Arambourgiania philadelphiae*. This giant animal had a wingspan of 39 ft (12 m).

Like *Pteranodon*, these giant flying reptiles fed by dipping their lower jaws into the water and scooping up fish that were swimming on the surface. The way the neck vertebrae interlock suggests that they had straight and rigid necks, which could be lowered from a hinge joint at the shoulders like the arm of an earth-moving digger. The great length of the neck is thought to have been necessary to prevent the enormous wings from hitting the waves while flying.

HOW THE PTEROSAURS FLEW

Generally, pterosaurs, with their short bodies and legs and relatively large wings, are thought – like bats – to have been clumsy when moving around on the ground. And, being reptiles, it was also assumed that they must have been ectotherms (cold-blooded) and so relatively inactive until their bodies were warmed by the sun, and incapable of sustained rigorous activity. Because of this, it was thought, they were probably not very active or efficient flapping fliers but gliders and soarers that sought out rising thermals of hot air to give them lift and height.

A remarkable discovery in Russia of fossilised hair and fibres associated with the pterosaur wing membrane has now challenged these assumptions. As the primary function of body hair in animals is to help them keep warm, it is possible that the pterosaurs, like birds, were warm-blooded and so more active than previously thought.

Even for the larger pterosaurs, takeoff speed was remarkably slow, possibly as low as 13 ft (4 m) per second. This speed could be reached after a short run into a head-wind, but as yet it is unclear how efficient pterosaurs were at running or walking.

PTEROSAUR LIFESTYLES

Questions as to how and where pterosaurs lived have been partly answered by a unique fossil deposit found in northern Chile. Remote but dramatically beautiful, most of the Early Cretaceous sediments in this harsh, semiarid part of the Andes are featureless desert sandstones laid down 130 million years ago. But on the slopes of the 13 000 ft (3962 m) Mount Cerro la Isla, a distinctive thick layer of conglomerate 6½ ft (2 m) deep was found. Such a deposit of pebbles and sand often indicates that there has been some dramatic change in

the normal sequence of sedimentation all those years ago, and so it turned out to be. The deposit stretched over ⅓ sq mile (1 km²) and clearly contained not just thousands but hundreds of thousands, if not millions, of bone fragments. Many of the fragments were thin-walled tubes of bone, and unmistakably the distinctive hollow and light bones from a pterosaur wing skeleton. The deposit was effectively a pterosaur graveyard, containing the jumbled remains of thousands of the creatures.

The sedimentary characteristics of the deposit suggested that it had been produced by a catastrophic flash flood, as can happen even today. An Early Cretaceous freak storm must have hit the mountains of northern Chile 130 million years ago, unleashing torrents of water. Hundreds of rivulets flowing down the hillsides collected into streams, and merged in the valleys into

a rumbling river of water and sediment. Always draining downwards, the flood was inexorably directed towards the lowest part of the local topography, the trough-shaped valley floor that now is 'fossilised' in Mount Cerro la Isla.

The size, thinness and structure of the hollow limb bones indicated that they belonged to immature pterosaurs of a single species, and that they belonged to animals with a wingspan of about 6½ ft (2 m).

Traditionally, it was thought that as pterosaurs had difficulty in running and beating their wings fast enough to provide enough lift for takeoff, they could only have taken to the air by flinging themselves off cliffs. The Chilean discovery challenged this view, as experts were able to build up a picture of behaviour from their find. The remains of so many thousands of bones of animals within a single flood deposit shows

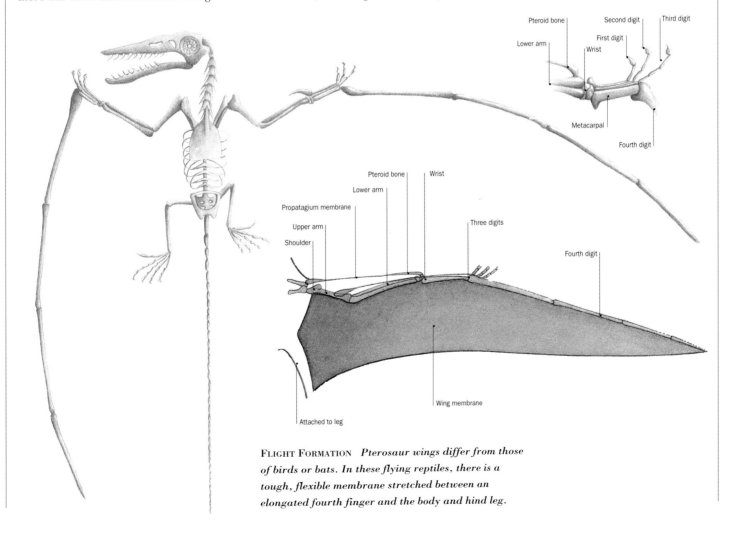

FLIGHT FORMATION *Pterosaur wings differ from those of birds or bats. In these flying reptiles, there is a tough, flexible membrane stretched between an elongated fourth finger and the body and hind leg.*

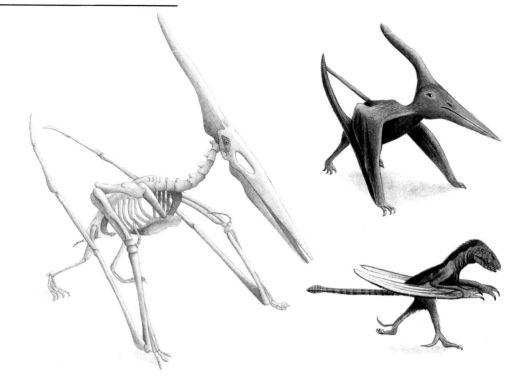

STRUTTING THEIR STUFF
Exactly how pterosaurs walked is debatable. Fossil trackways show that some pterosaurs, such as the Late Cretaceous Pteranodon (examples far left and top left) – with their elongated skulls and wingspans over 20 ft (6 m) – walked with both their feet and their hands. Others, such as the Early Jurassic Dimorphodon, may have folded their – considerably smaller – wings and run on their hind legs, using their tail as a counterbalance (left).

that they must have been congregated together. The implication is that the pterosaurs were living in some sort of colony when they were overtaken by the flood waters, drowned and buried.

This evidence of such social behaviour for these reptiles was a first, as they were previously thought to have been solitary animals (based on earlier isolated finds). This graveyard suggested they were more bird-like, with some species at least congregating far inland for mating and rearing their young. Furthermore, the adults must have been able to take off and land without the need for cliffs as launching pads to have lived in such an environment. But if they could get into the air so easily, why were so many drowned when they could have escaped the oncoming flood waters by taking off? According to scientists, the answer probably lies in the immaturity of the pterosaurs. Those that were overwhelmed

by the flood were hatchlings in a breeding colony or rookery, and were incapable of taking flight when the waters hit them.

A semiarid desert, surrounded by volcanoes and at least 12 miles (20 km) from the sea, seems an inhospitable place to build a rookery. But if the pterosaurs were powerful fliers, capable of 'surfing' on thermals with energy-efficient, soaring flight, they could easily have ranged far and wide in the search for food. The adults could have flown to the sea, caught fish and returned to feed their young at least once a day. And

raising their chicks in such a place would have provided protection from land-based predators. A similar breeding strategy is still used today by the grey gull in Chile, which establishes large rookeries within the semiarid valleys of the Andes up to 62 miles (100 km) inland from its fishing grounds.

This astonishing find provided the first solid evidence that pterosaurs had a bird-like group structure and system of parental care that was similar to that of their distant dinosaur relatives. It also opened up the prospect of an even more exciting discovery

TWO TEETH *The bone framework lightens the skull and jaws of Dimorphodon, whose name actually means 'two kinds of teeth' (right). Facing page: Some pterosaurs may have nested and bred in vast colonies, far from their coastal feeding grounds.*

one day in the future – that of an intact or abandoned fossil-breeding site lying deep within the ancient desert sandstones.

BIRDS OF A FEATHER

It was not long before the pterosaurs had to face competition from new rivals, who were to take over the airways of the Earth almost completely in the long run. Birds first appeared in the Late Jurassic era, 150 million years ago, but it was not until Tertiary times and the extinction of the pterosaurs that they really became established.

While the pterosaurs were still the leading animals of the airways during the Mesozoic era, a new evolutionary development was quietly taking place. Some small dinosaurs had evolved an unusual kind of body scale made of keratin, the protein of claws. Light and equally effective at protecting the animal against cold nights and hot days, these scales were actually the first feathers. The development of feathers is regarded as a defining characteristic of the successful group of egg-laying reptile-like animals we now know as birds.

In 1860, one of the most significant fossil finds in the history of palaeontology was discovered in Solnhofen in Bavaria. It was a single brownish-black feather, only 2 in (5 cm) long, flattened onto the surface of the rock. The rocks it came from were Late Jurassic limestone, 150 million years old.

As feathers are unique to birds, this discovery shocked the scientific community of 1860 – the only other known bird fossils were much younger. Here was evidence that birds were the contemporaries of dinosaurs, who themselves were just beginning to make an impact both scientifically and in the popular imagination.

THE FIRST EVIDENCE OF EVOLUTION

Fortunately for evolutionary studies, and for the development of palaeontology as a science, the find of the feather was followed the next year by that of a virtually complete skeleton of a bird. The fossil, which was also found at Solnhofen, was the general shape of a small theropod dinosaur, but showed clearly preserved wing and tail feather impressions in the rock.

The specimen was bought in 1862 by Sir Richard Owen, who at that time was in charge of the natural history collections at the British Museum in London. Owen was one of the great anatomists of the day, but he was implacably opposed to Charles Darwin's evolutionary ideas – which had first been published in extended book form in 1859. In 1863, Owen expounded his theory about this strange new specimen, known as *Archaeopteryx* (which is still amongst the

THEN AS NOW *Deep in the Chilean Andes, the valleys probably look the same today as when they were home to pterosaur colonies.*

A LITTLE BIT OF EVOLUTIONARY THEORY

The single specimen of the small bipedal theropod dinosaur *Compsognathus*, which was found in the Upper Jurassic limestones of the Solnhofen area in the late 1850s, is remarkable for both historical and biological reasons. It is the smallest known adult dinosaur, with the Solnhofen specimen being only 27$^{1}/_{2}$ in (70 cm) in total length. And, being entombed in calcareous mud, it is well preserved and complete. Once discovered, *Compsognathus* became one of the best known dinosaurs because it was used by Thomas Henry Huxley to illustrate the link between reptiles and birds in arguing the case for evolution.

Re-examination of the fossil remains in recent years has turned up even more remarkable features. In 1978, research showed that it was a carnivorous predator, as indicated by the fossil remains of a partially digested small lizard, *Bavarisaurus macrodactylus*, inside the body cavity. More recently, it has been proved that the specimen was an adult female – because of the presence of a dozen or so eggs that

have been partially preserved in the body cavity and the surrounding sediment. The small spherical eggs are just $^{1}/_{3}$ in (10 mm) long, partially flattened and show no sign of shell material. The shell is laid down just before an egg is laid, so its absence indicates that they were at an early stage of development within the body. As the carcass decayed, the intestines and oviduct, containing the string of immature eggs, would have burst from the body cavity and scattered the eggs onto the surrounding sea floor. There, they were covered with sediment and became flattened during further decay and later compression of the mud.

Another fine specimen of *Compsognathus* was subsequently found in a similar type of limestone from the south of France. Interestingly, the Paris specimen (as it is now known) is, at 5 ft (1.5 m), twice the size of the Bavarian one. It has been argued that this may indicate sexual dimorphism – meaning in this case that the males were significantly larger than the

females. This sort of difference between the sexes has been suspected to occur in dinosaurs but is very difficult to prove.

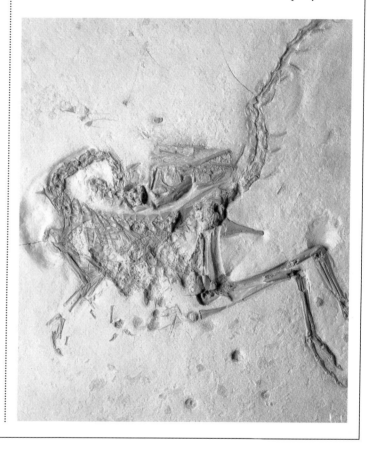

HUXLEY'S LINK *This chicken-sized dinosaur, preserved here with her eggs, is structurally similar to* Archaeopteryx.

most valuable fossils in the world and one of the Natural History Museum's most treasured possessions). He concluded that while *Archaeopteryx* was 'unequivocally a bird', it also had certain primitive characters which were not seen in adult living birds but can be found in the development of living bird embryos. It was later shown that all vertebrates show certain apparently primitive and common features – for instance, a tail – in their embryological development, which reflects a common ancestry.

The young zoologist Thomas Henry Huxley, a former protégé of Owen but now a great champion of Darwin and evolution, suggested that *Archaeopteryx*'s toothed jaws, long tail and 'embryonic' characters could

be thought of as simple reptilian features in what was, he agreed, more or less a true bird. Huxley also noted that a new Bavarian specimen of the small, bipedal dinosaur *Compsognathus*, found the same year as the feather, also had a number of remarkably bird-like characteristics and went some way to filling the gap between reptile and bird.

In 1868, in a lecture to a general audience at the Royal Institution in London, Huxley used both *Archaeopteryx* and *Compsognathus* to argue for the plausibility of evolution. These two fossils showed that, with evolutionary theory, there was no inherent problem in linking separate classes of animals – even with their apparently very different anatomy and physiology. Once

one common ancestor could be found in the fossil record, there was no reason why the other gaps should not be bridged.

There is little doubt that a great deal of the growing acceptance of Darwin's revolutionary ideas was due to Huxley's powerful advocacy, proselytisation and published essays. Huxley was a particularly persuasive speaker – in a way that the shy and private Darwin was not – and was able to reach wide audiences through his public lectures.

Now, palaeontologists realise that a fossil specimen of *Archaeopteryx* had already in fact been found in 1855. But it was not recognised as such until 1970, when it was spotted by the American scientist John Ostrom. The fossil had been languishing in a

FLIGHT FANTASTIC *The discovery, in 1860, of an* Archaeopteryx *feather was the first evidence that birds existed as long ago as Jurassic times.*

Archaeopteryx or 'ancient wing' – had down feathers, which proves that this oldest of birds was warm-blooded.

Nonetheless, the fossils found do suggest that the feathers were used for flying, even though self-propelled flight is not easy – as humans have found. It requires very well-developed flight muscles attached to a strong frame, a not-too-heavy body, large yet light wings of special design as well as the ability to build up enough speed to take off.

There are various theories as to how *Archaeopteryx* got off the ground. One is that the creature was fundamentally a small, two-legged and running theropod dinosaur, which had feathers for insulation and used its feather-covered arms as a kind of insect-catching system – as do some modern ground-dwelling birds. The wings also helped the animal leap into the air in the pursuit of its insect prey, and subsequently

museum in Haarlem in the Netherlands, labelled as a dinosaur. Yet, as Ostrom discovered to his amazement, when the fossil is viewed in the right light the faint impressions of feathers can be seen quite clearly. Since then, a further four specimens have been found (1877, 1951, 1955 and 1987), and from these a great deal more information has been obtained.

THE FIRST BIRD

Archaeopteryx was a medium-sized bird much like a European magpie, measuring about 11-19 in (28-48 cm) long from the tip of its snout to the end of its long tail. It was about 9½ in (24 cm) tall, and its light skull with large eyes and optic lobes in the brain show that it depended on sight as a key sense for survival. The narrow, pointed, beak-like jaws were armed with widely spaced, sharp teeth, while a curved neck led into a short back and on to a long, straight, bony tail made up of 22 vertebrae.

The bird's forelimbs had three greatly elongated fingers, each ending in a long curved claw. The hind limbs were particularly reptile-like, with the inner toe being very short and lying at the rear of the foot. This condition is also typical of many living birds, however, and, as Huxley pointed out, the foot of a chicken embryo is hard to distinguish from that of a reptile.

ON THE WING *The most common flying reptile of the Solnhofen sea,* Pterodactylus kochi, *was about the same size as a duck (facing page).*

The pelvis is also similar to that of a small theropod reptile, although it is still unclear exactly how it was constructed.

ARCHAEOPTERYX RELATIONSHIPS

It is unclear who *Archaeopteryx*'s closest reptile relations are, but the possibilities range from the crocodiles to the thecodont reptiles, and the mammals to the dinosaurs. There are certain similarities in skull structure with the crocodiles, although these are difficult to match exactly, as are those with the thecodont reptiles of the Triassic era. The mammal link is much more convincing, and is based on both groups of animals being warm-blooded and featuring four-chambered hearts and advanced brains. *Archaeopteryx* also had insulation made from the protein keratin, in the form of bird feathers and mammalian hair.

The strongest link, however, is with the dinosaurs. There are dozens of similarities between the skeletons of *Archaeopteryx* and advanced theropod dinosaurs such as the two-legged *Deinonychus*. Indeed, it is not surprising that *Archaeopteryx* has been mistaken for a dinosaur in the past – without its feathers, it virtually is a dinosaur.

THE DEVELOPMENT OF FLIGHT

The possession of feathers does not necessarily prove that *Archaeopteryx* flew. There are plenty of flightless feathered birds in existence today, and there is also evidence that one of the primary functions of the feathers may have been to regulate the bird's temperature rather than to fly. The

EVOLUTION'S ADVOCATE
Thomas Henry Huxley argued that Archaeopteryx *linked the reptiles to the birds.*

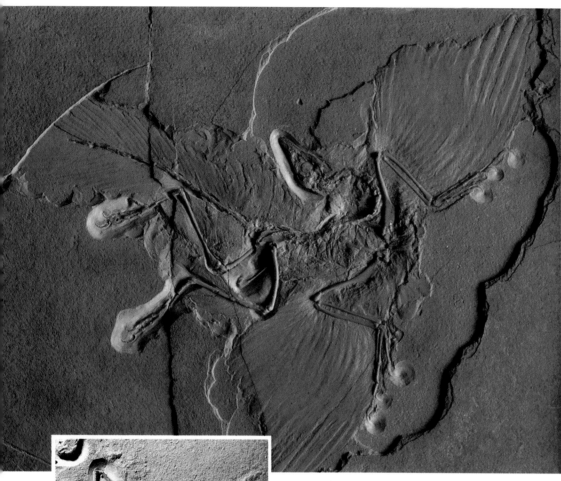

FOSSIL FAME *The best preserved* Archaeopteryx, *the 1877 Berlin specimen (above), is virtually complete. The earlier London fossil, with perfect tail feathers (left), was the first described.*

the aerodynamic demands of flapping flight. Birds that are flightless are characterised by flatter, symmetrical feathers.

An elevated launch site is another prerequisite for gliding – even on the shortest flights. Although *Archaeopteryx* has often been portrayed by some scientists as a forest inhabitant, capable of climbing trees with the hooked claws on its wing fingers and feet, the plant fossils recovered from the environment in which the bird lived do not include any trees. The only land available seems to have been low-lying islands in lagoons, which would have been covered by a sparse, low scrub of bushes separated by open plain.

With the prevailing arid climate, these islands would not have been able to support tree-sized vegetation. Some small, shrubby conifers such as the *Brachyphyllum* and *Palaeocyparis* and the bennettitalean *Bennettitales* were the tallest vegetation, and only grew to about 9½ft (3 m).

Nevertheless, there is little doubt that *Archaeopteryx* was capable of climbing. Detailed work on the wing claws has shown that their structure most closely resembles that of tree-climbing animals, rather than that of predators. For example, fledglings of a living South American bird, the hoatzin (*Opisthocomus hoazin*), have particularly well-developed claws on the first and second digits of their wings. The hoatzin nests in trees along the edge of waterways and the fledglings are vulnerable to attack from tree-climbing snakes. They escape by jumping into the water, but then climb back into the nest using the hooked claws on their wings. The claws are lost when they become adults and can fly.

Archaeopteryx wing claws face in a different direction to those of the hoatzin and cannot have been used in quite the same way but, nevertheless, they were sharp,

developed into powered flapping flight. A more plausible suggestion is that its flight developed from the ability to glide, which is a halfway step to flying but cuts out the need for flight muscles. Gliding could then have evolved more easily into powered flapping flight, which would have extended the animal's range.

Most successful animal gliders have developed in forests, since trees provide the necessary height (as long as the animal can climb). Being able to glide from one tree canopy to another has several advantages, as contemporary forest gliders such as

squirrels, monkeys and some frogs and lizards know well. It saves energy and reduces the risk of running into certain predators by not having to return to the ground quite so frequently.

Successful gliding also requires a tough wing membrane, which cannot be torn but which can be folded out of the way. Feathers are the ideal material, since they are light, waterproof and they do not tear easily. The feathers of *Archaeopteryx* are typical of flying birds in that they are asymmetrical and curved when looked at in cross-section. This particular shape is a direct response to

well-developed hooks, larger than the equivalent claws on the feet. They would have been a considerable help in climbing up and through vegetation, as well as for grooming, defence and perhaps hunting.

So *Archaeopteryx* was capable of climbing and flying, even though it lacked the keeled breastbone modern birds have for the attachment of flight muscles. Nonetheless, it was probably as good a flier as bats are today – who share the absence of a keeled breastbone – and its flight muscles could well have been attached to its strong wishbone. It does show various primitive features that link it with the reptiles, but it is a fairly advanced type of flying animal. Unfortunately, it is unlikely that any older, more primitive bird fossils will be found to fill in the missing links between reptiles, dinosaurs and birds.

AFTER *ARCHAEOPTERYX*

For most of the 20th century, the fossil record of birds has been non-existent between the Late Jurassic *Archaeopteryx* and the Late Cretaceous toothed birds, such as

TAKE OFF? It has been suggested that wing claws enabled Archaeopteryx *to climb into trees and use them as launch pads for flight.*

Hesperornis and *Ichthyornis*, found in the fine-grained limestone deposits of the Niobra Chalk Formation in Kansas. As a result of this, the link between these species and the majority of living birds – which are known as neognaths and are characterised by certain features of the palate and ankle structure – has been unclear. Not only do neognaths have an enormous diversity of form but they also have a generally poor fossil record.

Hesperornis was a flightless diver which stood over 3 ft (1 m) tall and looked rather like a penguin. It had a long neck, a smaller tail than *Archaeopteryx* and long, powerful back legs for swimming. The forelimbs were reduced to tiny, stick-like stumps that may have helped the bird steer while diving for fish, and its small, sharp teeth were probably designed to hold on to its slippery prey.

Ichthyornis was a smaller, gull-sized bird, with a large head and massive tooth-lined jaws. It was also a fish-eater, and probably caught its prey by diving while on the wing. Traditionally, it was thought that the evolution of birds was a gradual one, from *Archaeopteryx* through the Late Cretaceous aquatic birds such as *Hesperornis* to at least some of the modern Tertiary lineages. The scrappy fossil evidence indicated an unsatisfactory link between an extraordinary flood of new bird species during the Tertiary era which stemmed in some way from Mesozoic ancestors.

This picture was radically altered in the mid-1990s by new discoveries of Early Cretaceous fossil remains, which suggested that the birds had also suffered a massive extinction at the end of the Cretaceous period, along with the dinosaurs.

THE BOTTLENECK

According to this revolutionary new theory, which had actually first surfaced in the early 1980s, the few surviving birds acted as

---

STEPS TOWARDS HUMAN FLIGHT

There was a great renewal of interest in the theoretical origins of flight following the discovery of the *Archaeopteryx* skeleton in 1861, especially in North America. The idea that powered flight developed directly from the evolution of muscular, flapping wings on the ground was proposed by S.W. Williston in 1879. A year later, O.C. Marsh claimed that flapping flight had developed by creatures gliding down from trees.

The development of these theories and the improvement in understanding of the physics of flight increased the chances of humans becoming airborne in the 1880s. By the 1890s, the Lilienthal brothers of Germany made pioneering advances – both in theory and in practice.

When Otto Lilienthal died in a glider crash in 1896, his brother Gustav carried on alone.

FLYING COLOURS In 1891 the Lilienthals used pterosaur-like wings for their aircraft.

an evolutionary bottleneck – out of which all subsequent modern bird groups burst in Tertiary times. This concept came with the realisation that the majority of Mesozoic land birds could not have been ancestors of any of the modern birds, as was previously thought. Instead, they have been shown to belong to the enantiornithine, or 'opposite' birds, as they are called.

These important birds, which flourished and then became extinct in the Late Cretaceous period, were characterised by the particular way in which their ankle bones fused during development. This differs significantly from the way the ankles fuse in modern – or neognath – birds.

The Late Cretaceous aquatic birds, which included *Hesperornis* and *Ichthyornis* are grouped in a separate lineage, the ornithurines. This group coexisted with the 'opposite' bird forms, and fossil discoveries in the late 1980s from Early Cretaceous rocks in China (*Sinornis*), Spain (*Iberomesornis*) and Mongolia (*Ambiortus*) showed that the division between the enantiornithines and ornithurines was a very ancient one. Neither group managed to survive the Cretaceous/Tertiary extinction event.

Re-examination of the few Late Cretaceous fossil representatives of modern bird groups showed that they could all be characterised as 'transitional shore birds'. This meant that they could be linked to duck, flamingo and ibis-like shore-bird fossils that had been found in Early Tertiary rocks. As a result, scientists have concluded that all the modern bird groups, 9000 species grouped into more than 155 families, are derived from this small group of surviving shore birds – which were neither ornithurines nor enantiornithines.

This is one of the most remarkable bursts of evolution ever known. To produce, within 5-10 million years, all the major groups of living birds from these few survivors is quite extraordinary – but not unique. A similar evolutionary bottleneck produced the whales from terrestrial ungulate mammals over about 10 million years.

AN INCOMPLETE HISTORY

Despite the diversity and success of modern birds, their fossil record is remarkably patchy. This is in part because of the light construction of bird skeletons, which are hard to preserve well – as are the feathers. However, by analysing the processes by which organisms are buried – known as taphonomy – palaeontologists can tell which environments are most likely to accumulate certain kinds of fossil. Birds, like insects and plants, are best preserved in the fine-grained clay deposits of lakes, estuaries, lagoons and inland seas, such as those that were infamously laid down at Solnhofen in Bavaria. Fortunately, Early Tertiary times turn out to be well represented by such deposits in the London clay of southern England, the Early Eocene Green River Formation of Wyoming, and the Eocene-Oligocene of Quercy in France.

Subsequent Miocene deposits in Europe record the sudden rise of the song birds, or passerines, which are now represented by nearly 60 per cent of living bird species, or some 5700 kinds.

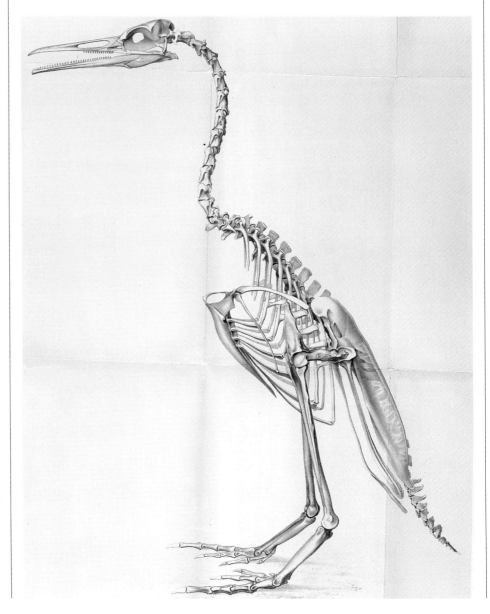

SLIPPERY FISH **Hesperornis** **aegalis,** *a common ornithurine* *bird, had teeth to help it hang* *onto its prey. It did not survive* *the Late Cretaceous extinction.*

THE END OF THE STORY?

SHEEP MENTALITY *Mammals such as Merycoidodon roamed the plains in vast herds.*

FLOWER POWER AND FUR COATS BECAME THE UNLIKELY SYMBOLS OF LIFE'S THIRD AGE — AND THE FINAL CHAPTER BEFORE MAN CHOSE TO MAKE HIS DEBUT. EXTINCTION HAD ALREADY REMOVED THE SUFFOCATING TYRANNY OF DINOSAUR DOMINATION, LEAVING THE WAY WIDE OPEN FOR EXOTIC NEW LIFE FORMS. FLOWERING PLANTS NOT ONLY COLOURED THE WORLD BUT ALSO PROVIDED A RICH, VARIED FOOD SOURCE TO SUPPORT THE NEW DIVERSITY OF SPECIES. THE ACQUISITION OF FEATHERS SENT BIRDS SOARING INTO THE HEAVENS. AT THE SAME TIME, THOSE WARM-BLOODED AND HAIRY CREATURES — THE MAMMALS — PERFECTED A STRATEGY FOR SURVIVAL THAT WOULD ENSURE THEY WOULD ULTIMATELY INHERIT THE EARTH.

NEW LEAVES *This specimen is from a Tertiary hazel tree.*

THE PUZZLE OF
THE GREAT EXTINCTION

Fire, flood, famine and drought are an unwelcome reality in the world today. But evidence from the distant past points to unimaginable catastrophes that literally wiped out whole species across the globe.

The 19th century was a crucial time in the development of science generally, and of geology in particular. In the early decades, there was a rapid growth in geological knowledge and in the techniques of investigation. By the 1850s, questions about both evolution and extinction were being considered.

Darwin's theories on evolution first appeared in print in 1859, and were in direct conflict with the literal interpretation of the story of creation. The idea of extinction was just as hard to confirm. Not enough was known about the living world to be positive that monster-sized lizards and mammoths were creatures of the past; they could still have been living somewhere on Earth.

WAS THERE A GREAT FLOOD?

Another question concerning geologists was exactly how the deep, wide valleys dissecting upland Britain and other parts of northern Europe had been formed. It seemed obvious that they had been cut out of the rocks by some power significantly greater than the rivers that flowed through them at that time.

As Christian beliefs were still firmly held by most people, it seemed logical to assume that this scarring of the Earth had in some way been caused by Noah's Flood. The Old Testament story of the Flood portrayed a wicked and corrupt Earth which

ICE SCULPTURE *It was not until the 19th century that geologists realised that upland Britain had been carved by glaciers.*

TERMINAL VIEW *The melting snout of the Kaskawulsh Glacier in the Yukon evokes the landscapes prevalent at the end of the Ice Age.*

was all but destroyed by a universal deluge sent by God. The sole survivors of this mass extinction were one pair of each kind of animal, saved by Noah along with his family in the wooden Ark he was instructed to build by God. Once the flood waters receded, both humans and animals began to construct a new life and repopulate the Earth.

This ancient story is remarkably powerful, and its retelling over 2000 years or so has led to a situation where the idea of global catastrophe is deeply rooted in the culture of the Judaeo-Christian world. The science of geology was hugely influenced by the idea of a universal deluge until well into the last century but, because of the overwhelming physical nature and catastrophic effect of the Flood, it is perhaps the one Bible story whose historical truth can be tested. After all, there should be evidence of both the physical repercussions of the Flood and the drowned remains of its victims.

Over the previous few hundred years, any discoveries of strange large fossil skeletons – obviously unlike anything in existence at that time – were explained away as victims of the Flood. There seemed to be no other way in which to rationalise the fossil skeletons of peculiar land animals such as the 20 ft (6 m) giant ground sloth *Megatherium* found in deposits from flood waters. But geologists soon began to find other fossils of extinct organisms in layers of rock that were much deeper that those thought to have been laid down during Noah's time. This suggested that not only were the skeletons older than the Flood, but also that the animals had already become extinct long before it. These findings coincided with the realisation that many of the stories in the Old and New Testaments of the Bible were not necessarily to be taken completely literally. Many people went on to believe that the true story could be revealed by the 'testimony of the rocks' – and so geology itself became a mission that could be pursued with an almost religious zeal.

A GLOBAL CATASTROPHE

By the middle of the century, it had become clear that it was a different kind of catastrophe that had affected much of northern Europe. Towards the end of the last Ice Age, around 15 000 years ago, the glaciers that had sculpted the region's upland landscapes released vast quantities of flood water as they melted. Tons of rock, sand and mud were swept down the slopes of mountains and hills and dumped over

WATER LOGGED *Layers of rock strata deposited on ancient sea floors are a reminder of repeated flooding of the land by the sea.*

the lowlands by the meltwater and, in the process, many geological features were formed that were similar to what would have been created by a universal flood. It is now thought that the story of the Flood

ASHES TO ASHES *Live volcanoes, such as Mount St Helen's in Washington, USA, are proof of the Earth's continuing dynamism.*

may originate from immediately after the end of the Ice Age, when early modern humans would have continued to experience changes in sea levels and occasional floods over many generations.

Late 20th-century attempts to anticipate the effects of a nuclear holocaust and the possibility of a 'nuclear winter' have led to a resurgence of interest in the study of natural catastrophes. From scaling up the known effects of major volcanic eruptions, it has become clear that the ejection of vast quantities of rock dust and smoke into the upper part of the atmosphere – as would happen after a nuclear war – could lead to their circulation all around the Earth. This in turn could result in a reduction in sunlight over a prolonged period of time, disturbing the Earth's established weather patterns long enough to lead to a dramatic cooling of surface temperatures. These changes could occur with such rapidity that plant life in particular might be irreparably damaged – with a knock-on effect through the whole of the Earth's food chain.

In this way, the damage from a severe blast and consequent radiation could affect the whole planet, especially if a period of darkness – a nuclear winter – followed. This research could help to answer a question that troubles palaeontologists: Did all the last dinosaurs die out at about the same time in different parts of the Earth, or was their extinction a gradual process?

By the 1970s, it was known that not only the dinosaurs but also many of their distant relatives did not survive beyond the end of the Cretaceous period, 65 million years ago. To pin down the timing of the so-called 'K-T' extinction, when dinosaurs disappeared, it was essential to work out exactly when the Cretaceous era ended and the Tertiary era began. This was not possible until a new tool for refining the timescale was developed in the early years of the decade.

THE DATING GAME

There are two major problems that crop up with any geological and palaeontological investigation. One is how to assess accurately the date of the fossils that are discovered, and the other is how to correlate the various fossils and the deposits they are found in from different parts of the world.

The dating of rocks has been possible for quite a while but, even using radio-active measuring, there is the possibility of a margin of error that can be as wide as hundreds of thousands of years. Radioactive dating is possible only on igneous rock anyway, as it works on the molten elements (from which it is formed). Putting an age on most sedimentary strata – formed from mineral and organic elements – is difficult. Generally, it is assessed by the fossils each layer contains.

By assuming that new species evolved and then spread rapidly – in geological terms – it is possible to match and therefore date the same species and their containing sediments as they are found around the world. Most of the species that are found globally were marine creatures, although some earthbound fossils, such as pollen, can be sufficiently abundant to form a basis for correlation.

An important breakthrough came in the 1970s when scientists discovered that the magnetic fields of various rock layers could give a guide to their age. The Earth's magnetic polarity goes through periodic reversals, and these changes are recorded in the layers of rock that are laid down at the time. By testing a sample of sedimentary rock, for example, it is possible to find out whether or not it was deposited under conditions of normal polarity. By measuring the switches in polarity in successive layers, the resulting record provides a time chart of magnetism which is as easy to understand as a modern bar code. By matching bar codes from different places with similar fossils, a more precise global correlation can be achieved.

WHEN DID THE EXTINCTION TAKE PLACE?

Such techniques led to a startling discovery in 1973. Geologists working on sedimentary rocks spanning the Cretaceous/Tertiary boundary near Gubbio, northern Italy, were able to pin down the boundary to a single thin clay layer, just a few inches thick, which lay within predominantly limestone rocks. And it was because of the microscopic shells of amoeba-like animals known as foraminiferans that Italian micropalaeontologist Isabella Primoli-Silva and American geologist Walter Alvarez were able to stumble on this boundary layer.

By analysing and comparing the number of shells in, above, and below the clay, they realised that marine foraminiferans – like dinosaurs and ammonites – had been catastrophically affected by the extinction event. The almost total absence of planktonic (surface-living) species above the layer seemed to indicate that they were the worst affected, with possibly as many as

PIECE OF CAKE? *Sections of strata, as in this classic 'layer cake' formation in Wales, are dated and analysed to produce a geological history of Earth.*

90 per cent of their number killed off. This evidence of a dramatic extinction event was far more convincing than anything suggested by the dinosaurs' fossils.

Walter Alvarez took samples of the clay layer back to the University of California at Berkeley, where both he and his father, Luis Alvarez, were professors. Alvarez senior, a Nobel laureate in physics, was interested to see what it would reveal when tested with highly sophisticated analytical equipment. Part of the analysis involved checking on the amounts of the heavy metal elements platinum and iridium in the clay. Luis Alvarez knew that these elements are exceedingly rare on the surface of the Earth at the moment but that they are present in meteoric dust which constantly rains down from space at a known rate. By measuring the quantity of these elements in the clay, it would be simple to estimate the length of time it took for the clay to accumulate.

Much to Alvarez's surprise, the clay contained 30 times the amount of iridium found in the layers of limestone above and below. There had to be some explanation for its concentration. Similar clay layers

SINGLE-CELLED CLUES TO THE EXTINCTION

Foraminiferans – or forams – are single-celled, amoeba-like, aquatic microorganisms which live in a variety of fresh water and marine environments. Many have shells, which have been found throughout the fossil record from the Cambrian era to the present. Since forams are often especially abundant in the sea, fossils of their shells are common in some ancient sediments.

The widespread distribution, common occurrence and relatively rapid rates of evolution of the

SHELTERED TOMBS *These nummulite shells originally accumulated on a seabed 40 million years ago. Now they form the rock of the pyramids at El Gîza, Egypt.*

forams have made them particularly useful when comparing the dates of layers of rock the world over. The pyramids and the Sphinx at El Gîza in Egypt, for example, were built with 40-million-year-old Eocene limestones, which are full of a particularly large, disc-shaped foram known as a nummulite.

Detailed analysis of where the forams were found in the clay boundary layer has led them to becoming particularly important in the debate over the extinctions at the Cretaceous/Tertiary boundary. Analysis has shown that 70 per cent of all the planktonic (surface-living) species became extinct within about 6 in (15 cm) above and below the boundary, but only rare species – about 20 per cent – disappeared all at once. Furthermore, over half of

the extinct forms lived within a certain depth zone in the oceans called the thermocline, and were particularly sensitive to any changes within it. Some experts interpret their extinction as the result of long-term changes in ocean circulation, which were initiated long before the date of the boundary layer as a result of plate movements deep underground affecting the configurations of the continents. Such changes in ocean circulation patterns have been shown to cause extinctions in other marine organisms, such as bivalved molluscs and belemnite cephalopods.

TOMORROW'S WORLD *A mass of foraminiferans on the seabed can eventually form layers of limestone (above). The amoeba-like* Elphidium, *for example (below), inhabits networks of spiral tubes within its chambered shell.*

HOLE IN ONE   *The Earth is constantly bombarded by meteors, but only large ones leave telltale craters – such as this one in Western Australia.*

found across the boundary in Denmark and New Zealand were sampled, and gave the same, unexpectedly high readings of the elements. The conclusion that Luis Alvarez came to in 1979 was that a massive meteorite had hit the Earth about 65 million years ago and had released a huge quantity of platinum and iridium into the atmosphere. This was then distributed globally, and fell to earth as dust, or sediment.

Walter and Luis Alvarez quickly made a connection between the meteorite impact and the extinction of the foraminiferans, as well as the ammonites, dinosaurs and other organisms thought to have suffered significant losses around the time of the boundary layer. It is now thought that up to 50 per cent of all species on Earth were made extinct at that moment and during the nuclear winter that may have followed.

Most scientists now agree that there was a major impact on Earth 65 million years ago, and that this event is marked by the Cretaceous/Tertiary boundary. For quite a while, however, it seemed hard to see where such an impact could have taken place. Surely there would be scarring on the Earth's surface if this was the case?

THE POINT OF IMPACT

In 1990, the Chicxulub crater in the Yucatán peninsula, southern Mexico, emerged as a possible candidate. The crater itself and rings of material thrown up by the blast stretch about 106-186 miles (170-300 km) across, and hardened molten rocks from the impact have been dated at 64-65 million years old.

These details seem to fit in with the known dates of the extinction event. But what is still not clear is how a meteorite impact can have selectively killed off particular groups of organisms. Why were the planktonic foraminiferans so drastically affected while other planktonic organisms were not? The answer to this may lie in the theory that this puzzling extinction was gradual, with only rare and perhaps vulnerable species disappearing in the long run. Some of the marine extinctions may in fact have been caused by long-term oceanographic changes, and were accelerated, rather than caused, by the impact.

Perhaps a similar last-nail-in-the-coffin situation was responsible for the vertebrate extinctions. Why did dinosaurs, pterosaurs and ichthyosaurs become extinct while other reptiles, such as crocodiles, did not? Certainly, there is no mass grave of bones to be found at the boundary, so comparing fossil evidence between the Cretaceous and Tertiary periods is difficult. It may be that it was vulnerable groups – those already in the process of decline – that became extinct. However, there is also evidence that some groups, such as certain marine reptiles, were on the point of major expansion when they disappeared from Earth.

SPACE ODYSSEY   *A high nickel content suggests this minute crystal from the Cretaceous/ Tertiary boundary arrived on Earth with an asteroid.*

THE AGE OF FLOWERS

From grasses to poplars and orchids to laurels, flowering plants provide the framework on which life depends. Their relationship with insects is known to be unique, but their origins and history remain cloaked in mystery.

Fundamental to life on Earth, flowering plants are the most dominant of all types of vegetation in existence today. They comprise 250 000-300 000 species, and are divided into three major groups: the monocots, which include grasses, palms, sedges and orchids, and account for about 22 per cent of all flowering plants; the eudicots, ranging from roses to sunflowers, which account for 75 per cent of living species; and the magnolias, plus laurels, which are thought to be one of the most ancient of modern flower groups.

All the major groups of flowering plants, or angiosperms, had evolved by the beginning of the Tertiary period, 65 million years ago – including the grasses which provide the basic human grain foods of rice and maize. Since that date, the landscapes of the world have been covered with vegetation very similar to that of today, and fossil evidence suggests that this ties in with a rapid expansion by the modern mammals. However, the early evolution of both mammals and plants is a contentious issue as scientists debate exactly when, and from what, the modern flowering plants arose.

During early Cretaceous times, the zones we think of today as the tropics were largely semidry, or seasonally dry, while higher latitudes, such as those of southern England, were extremely hot and humid. The link between changing environments and the transition from seed plants (or gymnosperms) to the flowering variety has yet to be properly explored, but humid conditions would have allowed some kinds of vegetation, such as those with a weed-type (ie, tough and opportunistic) life

FLOWER POWER *Angiosperms colour the world in order to attract insects and so ensure propagation. They comprise: monocots, such as orchids (below left); eudicots, such as sunflowers (below centre); and primitive magnolias (below).*

STONE BOUQUET *Flowers are essentially ephemeral, fragile and rarely fossilised. These 48-million-year-old* Florissantia *are the exception.*

fossil record of the magnolias extends back to 120 million years ago, and their pollen record goes back even further. Traces of the distinctive pollen have been found in rocks laid down 130 million years ago. So ancient are these fossils that the magnolias were considered for a long time to be the ancestral group for the angiosperms, especially as they show the most primitive structures of all the flowering plants.

If magnolias and eudicots had evolved by the even earlier Jurassic and Triassic times, the chances are high that traces of their pollen would have been found. The record of at least part of these periods is good and, in the absence of evidence to the contrary, it is fairly safe to assume that they had not yet developed.

How magnolias evolved from previous groups is not yet clear. The confusion lies in the fact that the detailed structures of magnolias – especially their flowers – are very different from those of preangiosperm plants from which they were thought to

---

strategy, to develop. These plants, which combined flowers and primitive features, are now thought to have been among the earliest of the angiosperms.

The problem with tracing the fossil record of the angiosperms is that most of the characteristics of flowering plants, such as the delicate reproductive structures (the flowers), are unlikely to be preserved. A few features, such as the presence of stamens with two pairs of pollen sacs and a carpel enclosing the ovule, have a slightly better chance of survival – but only in very well preserved fossil material.

HOW POLLEN COUNTS

As the fossil record of modern flowering plants is quite sketchy, much of the information known today about the plants has been discovered by tracing the pollen records of the various groups. The global

distribution of fossil pollen confirms that the early diversification of the flowering plants took place in the tropics of the time.

Little is known about the Early Cretaceous beginnings of the monocots, partly because their delicate structure is hard to fossilise without damage. The eudicots, on the other hand, have a distinctive pollen form which has been found in rocks that date back 125 million years. The flower

FRAGILE TRACES *Reproductive structures (right) are less likely to survive than pollen – as in this Early Cretaceous angiosperm (far right).*

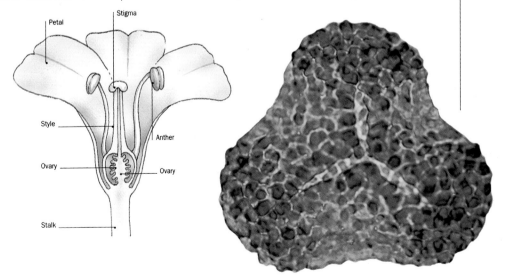

Petal

Stigma

Style

Anther

Ovary

Ovary

Stalk

FERNS BEFORE FLOWERS *Ferns
such as* Alethopteris *(facing
page) were the dominant
species until the flowering
plants took over in the
Cretaceous period.
Angiosperm ancestry may
originate as far back as the
Trias, with plants such as*
Pannaulika triassica *(right).*

have evolved including the
dominant cycad-type seed
plants (gymnosperms) of Jurassic times.

This lack of an obvious link between the
angiosperms and any of their possible
ancestors creates what is known as 'a morphological gap', a break in evolutionary history. In the case of angiosperms, the gap is
sufficiently wide to suggest that they must
have evolved from an older, and as yet unknown, group of fossil seed plants (unrelated to magnolias) in the Lower Triassic era,
more than 240 million years ago. If this is
the case, there remains an enormous time
gap during the Triassic and Jurassic eras –
nearly 100 million years –
through which the flowering
plants should have been evolving. The question of why there
should be no convincing flower
or pollen fossil record of this
time is particularly puzzling.

THE HIDDEN FLOWERS OF TRIASSIC TIMES

One theory, put forward in the
1990s, suggests that there is fossil evidence for the earliest
angiosperms which dates back
to the beginning of the morphological gap. Angiosperm-
like 'Crinopolles' pollen found
in Triassic sediments in North
Carolina dates back to around
220 million years ago, which

EARLY DAYS *Plants
such as this* Williamsonia
spectabilis *combined
cycad-like fronds with flower-
type reproductive organs.*

would extend the record of the flowering
plants back in time by nearly 100 million
years. The problem is that the parent plant
for this pollen is as yet unknown.

A possible candidate is *Pannaulika triassica*, a single fossil consisting of a small leaf
and two 'flowers' found in late Triassic (or
Carnian) shales. The pattern of veins in the
asymmetrical leaf fragment 1¼ in (3.2 cm)
long might suggest a typical early angiosperm. On the other hand, *Pannaulika*
also resembles certain ferns and, although
the fossilised plant material is well

preserved, the single critical leaf specimen
is incomplete. One answer to the dilemma
of the morphological gap may be that early
and as yet unknown Jurassic flowering
plants developed and evolved in tropical
upland regions. Plants that grow in this
kind of environment are generally unlikely
to be preserved as fossils, although their
wind-blown pollen can be carried beyond
the region of growth. Perhaps it is significant that the *Pannaulika* locality lay within
the tropics in Late Triassic times. The theory suggests that the flowering plants only
managed to spread out and down to the
subtropical lowlands, where they are found
as fossils, in Early Cretaceous times.

THE FIRST FOSSIL FLOWERS

Once established, new angiosperm groups
seem to have developed relatively slowly
until Mid to Late Cretaceous times, 30 million years later, when there is evidence of a
large number of new groups of flowering
plants. The evolutionary relationship between these groups is unclear, but a
comparison between the angiosperms and

an extinct group of Triassic plants – the *Bennettitales* – suggests that the evolving plant lineage that led to flowering plants must have diverged by Late Triassic times.

Some of the best preserved early flower fossils are those that were mummified in charcoal. Especially fine three-dimensional examples have been found in Mid and Late Cretaceous sediments from the eastern seaboard of America and western Portugal. Today, these areas are separated by the Atlantic, but 100 million years ago – when the ocean was just beginning to open – they were adjacent to each other. The most spectacular of these fossil flowers are from the appropriately named Rose Creek in Nebraska. This prodigious fossil site has turned up fossils of roses, buckthorns, saxifrages and magnolias, all of which show the five-petalled reproductive structures of modern bisexual, insect-pollinated flowers. Surprisingly, even the most delicate parts of the flowers are still intact.

HELPING HAND FROM THE ANIMALS

The diversification of the flowering plants could also owe a debt to nectar-collecting insects. Today, just as then, insects that feed on nectar are used as unwitting vehicles for transferring pollen from one flower's anthers (which contain the male reproductive gametes) to another's stigma (the female reproductive organ containing the ovules). This system of pollination is much more effective than the alternative of distribution by the wind, and allowed the flowering plants to colonise large areas quickly.

This insect/flower relationship cannot have happened overnight. Evidence suggests that some of the Jurassic seed plants, which had flower-like structures, were also pollinated by insects. Perhaps early angiosperms used the attraction of their flowers to bribe established insect pollinators away from their rivals, the seed plants.

Insects and other animals may also have played a role in dispersing the seeds and fruit of early flowering plants. The fruit of

---

SEEDS OF DOUBT: AN ANCIENT IMITATION

In 1993, American scientists found what turned out to be the only known fossil stick insect eggs. The tiny fossils, 1/8 in (3-4 mm) long, so closely resembled plant seeds that they fooled the palaeontologists initially. In so doing, they confirmed that stick insects have used mimicry to protect their eggs for at least 44 million years – in a reproductive strategy that is common among their modern counterparts.

Stick insects are a remarkable group of animals, with over 2500 species alive today. Some of these include the longest insects known to man, measuring 1 ft (30 cm) long. Their eggs are produced in large numbers but are relatively small, and more than 1000 contemporary species use plant seed mimicry to protect their eggs. In some cases, they are selected and buried by ants mistaking them for seeds – and this may also help to protect them.

The minute seed-like fossil eggs were found in Eocene sediments of the John Day Fossil Beds National Monument, in west-central Oregon. Once found, doubts about what exactly the eggs were led to them being sent to the British entomologist John Clark Sellick, a world authority on stick insect eggs.

Sellick's research showed that the fossil eggs belonged to three different species of stick insect, and are probably related to the living American genus *Anisomorpha*. These fossils are the first evidence that the mimicry is an ancient protective ploy, introduced to prevent the many small predators – from other insects to birds – from stealing the nutrient-rich eggs.

WALKING STICKS *The camouflage expert* **Phibalosoma phyllinum** *is just one of more than 2500 kinds of stick insect with the ability to mimic plant stems.*

DOUBLE ENTENDRE *Mating* **Anisomorpha buprestoides** *(above) produce huge quantities of eggs. Predators are a problem, but seed imitation (below) does afford some protection.*

RARE PRIZE *Fossil insects preserved in rock are relatively unusual. This Cretaceous hymenopteran was discovered in China (above). Right: The deal between plants and their insect pollinators involves the exchange of nectar for free pollen carriage.*

Mid Cretaceous period, 95 million years ago. This was the period that marked the change from the older ecosystems dominated by such plants as ferns, conifers and cycads to ecosystems that were dominated by angiosperms. For plants, this change was far more profound than that which occurred at the Cretaceous/Tertiary boundary.

Interestingly, there were changes in the plant-eating dinosaur communities somewhere around the Mid Cretaceous period, when the smaller, ornithopod dinosaurs took over from the large sauropods, who had long been the dominant dinosaur group. In fact, plant-eating dinosaurs may have been responsible for holding back the early development and diversification of the angiosperms, as it was not until the latter days of the dinosaurs that they spread out across the globe.

In Mid Cretaceous times, one of the older Mesozoic seed plant groups, the Gnetales, diversified in the tropics, although they later went into decline and were eclipsed by the angiosperms. Only three genera and about 40 species of the Gnetales survive today, all in semidry regions, while little held the angiosperms back. By the Mid Cretaceous era, angiosperms made up about half of the plant species on Earth. By the time of the Late Cretaceous era, they accounted for 80 per cent of all plants.

early angiosperms were much smaller than those of later counterparts. Modern flowering plants have relatively large seeds and fruit – designed to entice a range of animals, from birds and mammals (including humans more recently) to insects, and so ensure maximum distribution and propagation further afield.

Some modern plants use another device to aid propagation. They cover their fruit with small spines, which are designed to hook on to the fur of passing mammals until they drop or are brushed off. One plant that grew in the Early Cretaceous era shows that this may not be just a modern ploy, but most fossils found so far suggest that the first seed or fruit-eating animals appeared only in the Late Cretaceous era, becoming widespread in the Tertiary era. On the whole, however, the appearance of such animals seems to coincide with a time when angiosperm fruit and seed were increasing in size.

THE RISE TO DOMINANCE OF THE FLOWERING PLANTS

Wherever the flowering plants came from originally, fossil evidence shows overwhelmingly that they did not diversify until the

AGE-OLD DESIGN *Dissection of an 83-million-year-old* Scandianthus costatus *(below and left) shows adaptation for insect pollination.*

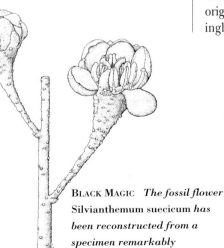

BLACK MAGIC *The fossil flower* Silvianthemum suecicum *has been reconstructed from a specimen remarkably preserved in charcoal.*

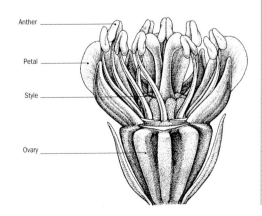

Anther

Petal

Style

Ovary

FOOD FOR THOUGHT *Tender plants, such as the Early Cretaceous* Bevhalstia pebja *(left), may have suffered from the attentions of grazing dinosaurs. By Tertiary times, flowering plants such as* Porana oeningensis *(right) had to face the onslaught of plant-eating mammals.*

Filling in some of the gaps in the history of the flowering plants would be easier if more of the gymnosperms (seed plants) that dominated the vegetation of the Earth during the heyday of the dinosaurs still survived. Only five gymnopserm groups exist today: the cycads, ginkgoes, Taxales, Gnetales and the conifers.

The fossil record of these gymnosperms has to be better understood for the evolution of the angiosperms to be resolved, but a combined molecular and morphological analysis now suggests that the earliest angiosperms were not necessarily large woody plants with magnolia-like flowers (as previously assumed). Instead, the earliest flowers may have been simple, small, fragile herbs. This would account for the lack of previous history for the angiosperms, as scientists would have been looking for the wrong kind of plant as a angiosperm ancestor. Also, small, fragile herbs are less likely to be preserved than a robust magnolia.

Fossil finds in the mid 1990s seemed to support this view of the early flowering plants. Palaeontologists working in the south of England had been alerted to watch out for vegetation and put anything plant-

like to one side. In Early Cretaceous times, the area had been a subtropical lowland, traversed by meandering rivers and covered with a lush vegetation still dominated by cycads. There was abundant animal life, including insects, fish and dinosaurs such as *Iguanodon*, *Baryonyx* and *Megalosaurus*, and it was felt that much could be revealed from Wealden clay in the area.

THE OLDEST FLOWER

Slabs of clay examined under a microscope, started to unveil numerous bits of stem and small, fragile leaves that looked as if they might belong to the same kind of plant. Once reassembled, they formed a small, fragile wetland herb that stood 10 in (25 cm) tall – the oldest fossil flower yet found and the missing piece of the puzzle.

The plant, *Bevhalstia pebja*, combines primitive features, such as a fern-like anatomy and leaves, with more advanced branching and flower-like structures. Its unique form is quite unlike any other plant from Early Cretaceous times, including the magnolias. Scientists continue to search for fossil

evidence of its pollen in surrounding rocks, but so far none has been found. If it is a true flower, the microscopic pollen grains will show a unique characteristic surface texture and so provide a vital clue to the history of the flowering plants. But while the discovery of *Bevhalstia* will revitalise the quest for angiosperm origins, it does not solve the outstanding problem of the wide morphological gap. The answer must lie buried somewhere in the Late Triassic or Jurassic rock record.

HISTORY OF LIFE *The* Ginkgo biloba *(right), originally native to China and cultivated in Chinese and Japanese temple gardens, is now found worldwide. It is truly a living fossil with a record extending over 200 million years. Facing page: A fossilised leaf of the* Ginkgo biloba.

LOOK BACK
IN AMBER

Once worth its weight in gold, amber is now almost priceless – but for scientific reasons. Not only does it perfectly entomb and preserve the most fragile species, but it is also proving a source of ancient DNA – the essence of life itself.

Amber is a curious, almost magical, substance which has been coveted and endowed with special worth throughout history – treasured for its unique translucence and depth of colour.

It first appeared in the distant geological past as a sticky resin that oozed from the earliest forest trees as a protection against fungal and bacterial infection. As it rolled down the outside of a tree, it entombed

A GOLDEN GRAVE *The hunting days of this Chrysopilus – an insect predator – ended about 38 million years ago.*

countless small insects in what became, in time, a clear, completely airtight surround. As a result, amber contains an unparalleled record of fossilised life.

Appreciation of amber as a decorative material, worth its weight in gold, came to a peak in 1701 when King Frederick I of Prussia commissioned an entire room made of amber as a gift for Peter the Great of Russia. It later went through a renaissance in Victorian times, when amber beads became

a popular adornment for matriarchal bosoms. It may not have the same decorative appeal now, but the fact that amber was prized and preserved throughout the centuries – amulets of amber have been found that date back as far as 35 000 BC – makes it a fossil treasure trove for palaeontologists today.

A LEAD FOR THE SCIENTISTS

Organisms suspended in the amber – whether ferns, insects or feathers – have always been part of the allure. Even in the 1st century AD, the Roman Marcus Valerius Martialis remarked that 'the bee is enclosed, and shines preserved in amber, so

CROWNING JEWELS *A preserved spider and a cricket add a hint of the macabre when combined with diamonds in this amber pendant.*

that it seems enshrined in its own nectar'. His countryman, Pliny, noted that amber was the discharge of a pine-like tree and often contained small insects.

It was not until the 19th century that any major collections of amber flora and fauna were started in earnest. The largest was amassed in the Baltics by an innkeeper, Wilhelm Stantien, and a merchant, Moritz Becker, who used dredging and mining operations to extract pieces of amber from the Samland peninsula, near the Baltic port of Königsberg (now Kaliningrad). In total, they extracted 120000 amber-embedded animal and plant fossils from Tertiary greensands and clays laid down roughly 38 million years ago. But, tragically, much of this amazing collection was lost during the Second World War. There are still large collections of Baltic amber in museums worldwide – the Natural History Museum in London has some 25000 specimens – but in total they do not amount to much more than that one unrepeatable collection.

WHERE DID THE AMBER COME FROM?

The history of amber resin-producing trees extends back to Cretaceous times, over 100 million years ago, and may date as far back as the late Carboniferous era, 300 million years ago. Worldwide, many different regions – from the Dominican Republic to China and Romania – produced amber in Tertiary times, and an astonishing amount has been recovered. The Baltic has proved a particularly rich source, as Stantien and Becker's operations proved. They alone produced between $1/4$-$1/2$ million tons of amber a year between 1875 and 1914, and probably well over 10 million tons in total. This led to a popular misconception about amber: that it was the fossilised resin of coniferous trees from the Baltic region, and that its abundance was the result of an unusual pathological condition of those ancient trees.

Modern research has debunked many of these myths. It is known now that amber was produced by different types of tree worldwide, and that the most likely candidate responsible for Baltic amber is an araucariacean similar to the living *Agathis* from New Zealand. The araucariaceans, evergreen trees which form part of the coniferous group, secrete huge amounts of a preservable resin that is brittle, aromatic, and yellow or reddish in colour. Given hundreds of thousands, if not millions, of years to accumulate, this 'copal' resin could easily have been responsible for the amount of amber found in the Baltic.

ON THE TRAIL OF INSECT DNA

Amber provides a uniquely well preserved view of the past for those interested in fossil insects. Insects are probably the most successful group of animals to have existed, and they can be both useful and devastating to humans. The more that can be understood about them the better and, at the moment, amber provides the only fossil context in which ancient insect relationships can be studied.

Fossil amber samples have revealed diverse information, from the fossil record of the Mycetophagidae (hairy fungus beetles) to the behaviour of parasitic wasp larvae towards their spider hosts. As the 18th-century satirist Jonathan Swift wrote in a famous ditty:

So naturalists observe, a flea
Has smaller fleas that on him prey;
And these have smaller still to bite 'em;
And so proceed ad infinitum.

MEDICINE DROPS *For millions of years, plants such as the Scots pine –* Pinus sylvestris *– have produced special resins to protect themselves against fungal attack.*

field of molecular palaeontology (and was also the premise behind Michael Crichton's mid-1990s blockbuster, *Jurassic Park*).

Before they turned to amber, scientists first attempted to extract DNA from fossilised remains in the early 1980s – but there were no significant results until 1984. Then, the scientific world was stunned with the announcement that DNA had been sequenced from the dried skin tissue of a quagga, an extinct zebra-like animal from Africa, and a 40 000-year-old mammoth. Further specimens varied from a 2400-year-old mummy to a 17-million-year-old magnolia leaf.

However, such research was not without its problems – or its detractors. First, a powerful amplification technique (PCR) had to be used to detect the complex giant molecules of DNA, which are fragile and rapidly degrade upon the death of a cell. The technique is so sensitive it can detect and sequence a single DNA molecule, but there are fears that it may also pick up on contamination in the sample. Some scientists say that it is impossible to sequence any DNA older than a few tens of thousands of years old. Anything older, they say, is unwitting contamination. The only way to test the accuracy would be to reproduce the results under similar conditions, but this has not yet been achieved. The second problem is that there are technical limitations with this type of fossil material – which, in any case, is not generally available.

But even more significant than the fossilised insects was the discovery that it was possible to extract deoxyribonucleic acid (DNA), the genetic make-up of a cell, from animals trapped in amber. This scientific breakthrough promised to open up the

For the dedicated few, the idea of obtaining information about fossil DNA and the importance of its potential applications were so great that they persisted with the search for better sources. What they were looking for was a situation in which organisms were virtually buried alive, sealed in a capsule that dehydrated the body and provided an antiseptic environment – and therefore excluded any microbes that would cause decay and leave confusing traces of their own DNA. This is a tall order for the natural environment, which usually promotes decay, but amber provides circumstances that are as near to the ideal as possible.

Its potential was realised in 1982 when scientists were able to perform remarkably detailed analysis of the body tissue of a fly found embedded in amber from the Dominican Republic. There, extensive mining has thrown up an extraordinary range of animal life in the amber, including such rarities as a frog, a mushroom and mammalian hair. DNA has now been extracted from insects that were trapped more than 30 million years ago – such as a termite *Mastotermes electrodominicus* and a stingless bee *Proplebeia dominicana*, both of which were found embedded in the Tertiary-age amber. The fossil termite has been shown to share several DNA sequence attributes with a living Australian termite long regarded as a 'missing link' between cockroaches and termites.

In 1993, a team of scientists took the concept back even further by managing to sequence the DNA from a 130-million-year-old weevil. This pushed the possibility of extracting at least part of the 'code of life' back to the time – if not the actual life – of the dinosaurs.

AMBER IN BRITAIN

Cretaceous amber 115 million years old has just been discovered for the first time in the British Isles. This unusually old fossil amber, of an age previously only known in amber from the Middle East, was found in 1993 in ancient river sediments exposed on the south coast of the Isle of Wight. The deposits containing these remarkable amber pebbles have been identified as part of the Wealden Marls, which date back to the Early Cretaceous period.

A reconstruction of life on the Wealden Marls all those years ago suggests that the amber resin was produced in forests which were dominated by cycads (seed plants) but which also contained some coniferopsids (conifers). The trees would have exuded the amber resin as a protection against the bacterial attack suffered by most plants, and its aroma would have acted much as a fly paper does today. Insects (such as wasps) that unwittingly landed on its sticky surface were doomed, and the more they struggled the more entangled they became. The resin hardened, then accumulated along with

continued on page 132

SELF-PRESERVATION *Tertiary resin from a* **Hymenaea** *tree in Dominica managed to entrap one of its own flowers, about 30 million years ago.*

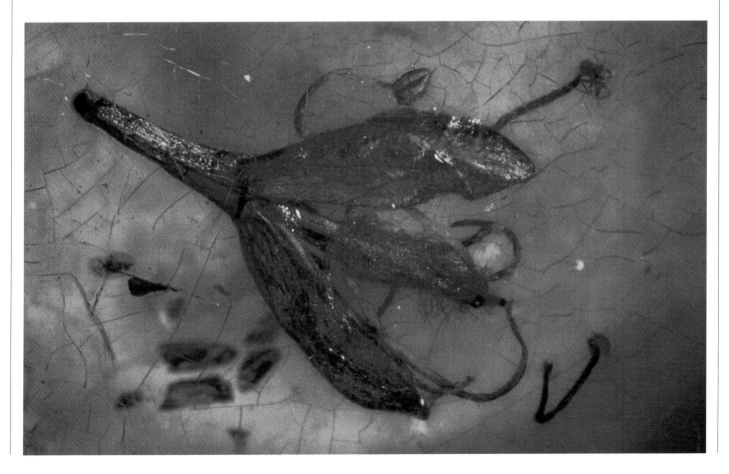

FOREVER AMBER One of the
best media for fossilisation,
amber preserves even the
delicate wings and legs of flies.

wood fragments to become part of the forest floor litter. Migrating river channels cut into this debris and carried it southwards until stagnant pools accumulated on the flood plain, trapping and preserving

HOW AMBER CAN TRAVEL

Prior to the discovery of Cretaceous amber in the south of Britain, most specimens were found along the North Sea coast. Essentially, this was Baltic amber, dumped on the floor of the North Sea among piles of debris transported by melting glaciers. Amber has a low density and often contains air bubbles, so it is quite buoyant in sea water. It can be carried considerable distances by currents before being washed up as flotsam on a coastline.

Amber that has been reworked by the sea and landed on the northern coasts is much younger than the Cretaceous amber found in the south of the country. It dates from Tertiary times, 38 million years ago.

the wood and amber. The corpses of small animals that had died and been swept downstream, such as that of a small bipedal dinosaur *Hypsilophodon foxii* or perhaps an

Iguanodon, rotted away, leaving their bones in the sand at the bottom of the pools. Dinosaur bones and footprints have been known from this area for some time, and small flowering plants such as *Bevhalstia pebja* might easily have bloomed for the first time in such a marshy region. Indeed, their remains have also been discovered in the Wealden Marls.

The Early Cretaceous Wealden sediments in which these discoveries were made are well exposed, not only in the cliffs of the Isle of Wight but also throughout Kent and Sussex on mainland Britain. Extensive geological research throughout the southeast of England over the last 100 years means that a great deal is already known about the Wealden environment, which is one of the most intensively studied layers of sedimentary rock in the world.

Interestingly, the Cretaceous amber was found with abundant lignite 'logs' – fragments of plant material which had fossilised into low-grade coal. These logs, together with plant remains, suggest that the amber

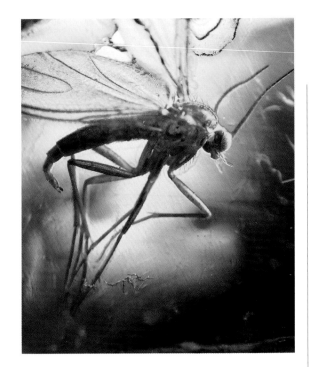

INHERITANCE TASKS
Fragments of genetic material have been recovered from insects such as this fossil midge from 30 million years ago.

originated as resin from *Brachyphyllum*, a conifer from the Araucariaceae family. One way of testing such a theory is by 'fingerprinting' with infra-red spectroscopy – a technique which uses chemical analysis to distinguish amber from other resins. The results show some similarities to Cretaceous ambers found in Canada and Israel, but lack other features thought to be characteristic of old ambers.

What they do suggest is that the amber originated not from an araucariacean tree but from another related family known as the Taxodiaceae. This apparent discrepancy between the evidence from the plant fossils and chemical analysis may be because in Early Cretaceous times conifers were still in the process of evolving.

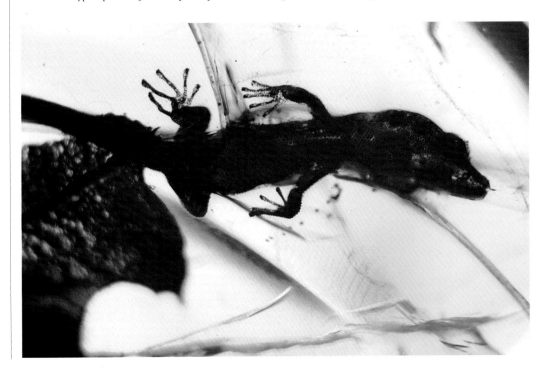

LUCKLESS LIZARD *Large animals are rarely entrapped by resin, but here a gecko from 40 million years ago has been caught up in Dominican amber.*

FLY-BY-NIGHT SPECIMEN THAT FOOLED THE EXPERTS

The value of insects trapped in amber has been sufficiently high to attract forgers for more than 100 years. Even experts and major museums can be taken in by the forgers' expertise, as the 1993 discovery of the entomological crime of the century shows.

One day, Andrew Ross, a palaeoentomologist at the Natural History Museum in London, was hard at work on a unique and almost perfectly preserved specimen of a latrine fly, *Fannia scalaris*. Preserved in amber and thought to have originated from the Baltic region 38 million years ago, it had acquired considerable scientific fame over the years. Not only was it the oldest known representative of the family Muscidae, but it was also an important example of a species that had lasted a remarkably long time. *Fannia scalaris* as a species is still very much with us.

As Ross sat hunched over his microscope in one of the research laboratories, there was a brilliant

INNOCENT VICTIM *Fannia scalaris is commonly known as the latrine fly.*

golden glow from the precious amber – illuminated in the bright white light of an intensity lamp. Suddenly, to his alarm, two cracks appeared on either side of the specimen. He quickly removed it from the heat of the lamp and, turning it around to check the extent of the damage, Ross's attention was drawn to a faint but clear line running through it. As he focused on the line, he noticed something even more curious – a hemispherical depression inside the amber, in which the specimen appeared to be sitting. In a flash, Ross realised the implications for the scientific world. This was a classic case of a masterfully executed doublet, a technique perfected by jewellers over the centuries to stretch their valuable stones.

A genuine piece of Baltic amber had been carefully cut in half and a small, grave-like excavation made in one surface.

Then, the specimen of *Fannia scalaris* had been placed in the centre of the depression and the cavity filled with an amber-like resinous mounting medium. Next, the two original pieces of amber had been glued back together again to form the doublet, and thus the fake was all ready to be launched on its – remarkably successful – scientific career.

The fake's point of entry into scientific stardom can be traced back to its acquisition by the eminent German dipterist – or student of flies – H.F. Loew. He mentioned the fossil fly in a published list of his collection in 1850. The specimen lay untroubled for more than 100 years until the world-famous entomologist Willi Hennig made a detailed study of it in 1966. This study firmly established the fly's international scientific reputation as a kind of muscid *Marie Celeste*, floating out of the distant past. The fly was clearly identifiable,

DOUBLET BLUFF *Viewed from the side in close up, the fly can be seen to be laying in a sort of excavated depression within the 'solid' piece of amber.*

and appeared to have remained unchanged by evolution for at least 38 million years – although it had no known forebears or descendants until the present.

Fannia scalaris was even included in a leading work on fossils, the *Treatise on Palaeontology*, where it was described as seeming 'to be identical with an existing muscid, *Fannia scalaris* Fabricius (1793)'.

It was probably the nearest insect to hand when the forger, whose identity will never be known, tried his or her hand at deceit. Probably the motive was money, although the forger also successfully bamboozled scientists for more than 140 years while the fake lay undetected in the Museum's collection.

EXTINCTION AND RADIATION

Waiting in the wings of the Mesozoic era, mammals and birds spent millions of years under the shadow of dinosaurs and their relatives. But, when the time was right, the young pretenders would stop at nothing short of world domination.

Massive global upheavals accompanied the breakup of the Pangaean supercontinent and land areas of Laurasia and Gondwanaland in the Early Cretaceous era, 130 million years ago, and this triggered off a complex chain of events. The ocean circulation patterns around the world changed, as did the climate and eventually the vegetation. This meant that all animal life on Earth was probably forced to adapt to new climatic conditions in order to survive. Up until then, the dinosaurs and their reptile relatives had dominated the land, air and sea but, soon after all the global upheavals, mammals and birds experienced a huge explosion in both numbers and species.

Mammals and birds had been quietly coexisting with the dinosaurs since Late Jurassic and Late Cretaceous times, about 150 and 85 million years ago respectively. But, as the dinosaurs later faded away, the ecological niches they had occupied were left empty. These were

INHERITING THE EARTH
Most large mammals are plant-eaters, and those such as springboks (above) are capable of surviving on the toughest desert shrubs. Over the last 65 million years, large mammals such as deer, giraffe and zebra (below left) have dominated land faunas.

quickly filled by the mammals and birds, which may well have been held back from spreading sooner by the dinosaurs' success – according to traditional theory. More recent thinking, however, suggests, that the environmental instability caused by the Early Cretaceous global upheaval had necessitated and initiated adaptation in the birds and mammals, who were perhaps able to exploit the new habitats to a greater extent than the dinosaurs.

The extraordinarily rapid diversification of mammals that followed must be one of the most remarkable events in the whole history of life. Vast numbers of plant-eating mammals developed in response to the evolution of the modern flowering plants in the Early Tertiary period, and these herbivorous animals diverged dramatically in their forms in order to exploit the new food opportunities. Within 65 million years, from the beginning of the Tertiary era to the present, the mammals evolved into nearly 4000 species, grouped into 129 families. Much of this expansion was concentrated into the 10 million years or so of the Palaeocene and Early Eocene eras. This period also saw the development of 15 or more new lineages – distinct groups of closely related organisms, which evolve together through time.

Diversity of vegetation was not the only reason for the mammals' success. Perhaps the most important event of their evolution was the ability developed by some groups to prolong the gestation of young within the body. This came with the development of a larger opening through the pelvis, an advance which allowed them to expand into new habitats. They could now give birth to precocious young that were already well developed, and needed less protection because they could be born when they were more ready to fend for themselves. This also encouraged the development of social behaviour both in families and in clans.

WHAT THE FOSSIL RECORD SAYS

The early fossil record of mammals and birds is sketchy, as neither group preserves well in sediment laid down on land – especially not the delicate bones. The only parts of the skeleton with much resistance to weathering, erosion or passage through the gut of a predator are the teeth of mammals, which can be difficult to find as fossils.

As there is so little fossil evidence, dating when exactly the expansion of the birds and mammals took place is still open to debate. One view links the expansion to the extinction event that led to the demise of the dinosaurs, while another uses the 'molecular clock' to assess how long ago the major living groups diverged from their common ancestors. This is based on known rates of genetic evolution and assumes that the greater the genetic distance between the groups, the longer ago they must have evolved into different lineages.

Using this scale of measurement, the major groups of living mammals and birds would seem to have originated between 50 and 90 per cent earlier than the extinction event, which ties in with the breakup of Pangaea. There is little fossil evidence – so far uncovered – to support this idea, except from the Gobi desert where spectacular finds do indeed seem to suggest that birds and mammals developed far earlier than was previously thought.

ACCURATELY IDENTIFYING A MAMMAL

Correctly identifying a fossil mammal is not all that simple a task. The characteristics that are shared by mammals – a heart with four chambers, body hair and mammary

TOOTH AND JAW *A jaw made up of a single bone and tricuspid teeth help to identify the Late Triassic Morgancucodon as one of the first mammals.*

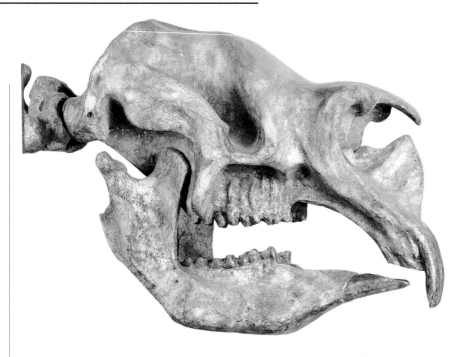

glands used by females to suckle their young – are all made of soft tissue which disintegrates in time and so is unlikely to be preserved in rock.

Fortunately, structural changes that gradually occurred to the skeleton while the mammals evolved can be seen in fossils. One of these is the dentary, a single bone element which has enlarged to make up the entire lower jawbone and which is shared by all living mammals. This modification changed the way the lower jaw articulated with the skull, while some alterations in the muscle structure of the jaw gave the animals improved biting and chewing skills, especially in the cheeks.

Another clue lies in the diminutive but all-important bones of the middle ear. These three bones – the malleus, incus and stapes – developed from bones found in the lower jaw of reptiles and are now crucial for all mammals. They help to transmit the vibrations collected by an external ear to the inner ear, so improving the animals' hearing and communication, which in turn led to developments in their social behaviour.

IN LINE *The shrew-sized* **Megazostrodon,** *a morganucodontid fossil, provides the best idea of the body form of early mammals.*

Tracing the story of mammalian evolution is far from simple, however, partly because of the geographical isolation of the great continents of South America and Australia. Today, there are three surviving groups of mammals – the monotremes, the marsupials and the main group of placental mammals. But while these groups do share a few features, they differ hugely when it comes to methods of reproduction.

The monotremes, which include the platypus and spiny anteater (Echidna) of the Australian continent, lay eggs, while the marsupials give birth to tiny, immature young who continue to develop in an external pouch and are attached to the mammary teat. Kangaroos belong to this group, as do the opossums of North America. By comparison, placental mammals retain embryos in the uterus far longer. There, they are nourished by the mother, kept at a con-

MIND THE GAP *Behind chisel-shaped incisors, the skull of a plant-eating* **Diprotodon** *had a large space for manipulation of its food by its tongue.*

stant body temperature and are born at a more mature stage of development.

All three of these mammalian groups were around in the early days after the great extinction, as were other groups such as the multituberculates (so named because of the many rounded cusps on their teeth) which subsequently became extinct.

THE FIRST TRUE MAMMALS

It seems that tiny, shrew-like creatures, whose fossils have been found as far afield as Europe, China and southern Africa, were the first true mammals. The oldest of these fossils is *Morganucodon*, whose skull was found in rocks first laid down in the Late Triassic era 210 million years ago. The fossil skull measures 1 in (2.5 cm) long, and shows distinct mammalian traits – such as an almost complete bony braincase, a lower jawbone with the dentary bone greatly enlarged, and well-differentiated cheek teeth.

A better understanding of these mammals came with a lucky find in southern Africa. A single, almost complete specimen of a *Megazostrodon* was found curled up in the rocks. It was almost small enough to fit into a matchbox but, when extended, its skeleton measured about 4 in (10 cm) long. *Megazostrodon* appears to be similar to a rodent, with its flexible backbone, long tail and delicate-looking limbs, but it has

important differences from its living counterparts (which did not evolve until much later in Early Tertiary times, more than 50 million years ago). The skull is large in relation to its body size and it has long jaws and complex mammalian teeth. Its forelimbs stuck out to the side, like those of a reptile, which would have given it a sprawling gait. *Megazostrodon* was probably more capable of brief, rapid motion than sustained running.

THE IMPORTANCE OF GOOD TEETH

Teeth serve more functions for warm-blooded animals, such as mammals, than for cold-blooded ones. Not only are they used for protection and to secure food, but they also help in predigestion. When food is chewed thoroughly before swallowing, it can be digested more easily and so releases an almost constant supply of energy. This is needed to maintain body temperature, as well as all the other business of life.

Both *Morganucodon* and *Megazostrodon* had long, sharp, stabbing, fang-like canines, which were separated at the front of the mouth by a row of small chisel-shaped incisors. Food was torn and chopped up before being passed back to the cheek teeth by the tongue. As the lower jaw was capable of moving back and forth and sideways, and the cheek teeth had complex ridges and cusps which fitted together, pieces of food could be chewed or shredded up before being swallowed.

Such teeth suggest that the animals must have had a carnivorous diet, although they themselves were so small that their prey must have been small invertebrates – such as insects. Nevertheless, these primitive mammals survived for more than 50 million years, during which time several other mammal family lineages – most of which have been identified by their teeth and fragments from their skulls – evolved.

OUT OF MONGOLIA

One of these lineages formed the largest group of early mammals, the extinct multituberculates. These had evolved by Late Jurassic times, 140 million years ago, and possibly much earlier. They survived the Cretaceous/Tertiary extinction event only to die out 35 million years ago.

It is sometimes hard to imagine animals from such a distant past, but a wealth of information about the multituberculates came to light in the Mongolian Gobi desert in the 1960s and 1970s. The skeletons of six different kinds of multituberculate mammal were uncovered in Late Cretaceous rocks, with the bones so well preserved that scientists were able to reconstruct exactly how the muscle structure of the animals was put together. From this, they were able to show that the multituberculates had very similar habits to certain small, living marsupials and rodents. Earlier finds of multituberculates, such as *Ptilodus* from the Early

PET ANCESTORS *At first, early mammals were considered tree climbers but now they are thought to have resembled gerbils (left). Gerbils burrow in treeless deserts such as those of Turkmenistan (below).*

BABY-BEARER *It is hard to prove when mammals started giving birth to advanced babies, as in the case of the elephant shrew.*

Tertiary period of Canada, had led to theories that they were tree climbers, but fresh information from the fossils found in Mongolia showed that the animals were in fact much closer to living gerbils (or jirds) in the way that they lived.

By working out exactly how the joints were operated, and by which muscles, scientists reconstructed the shoulder girdle of a multituberculate known as *Nemegtbaatar*. This suggested that the animal had a sprawling stance which was much better adapted for a particular mode of running than for tree climbing. It could probably produce steep jumps or hops over short distances, and bursts of speed with an asymmetric gait.

Animals that live in burrows, such as gerbils, need just such running skills to survive. While foraging for food, they keep a constant lookout for danger, occasionally dashing for the cover and safety provided

by their burrow when they sense a threat. Geological evidence from the sediments surrounding the Mongolian fossils suggests that these multituberculates lived in much the same environment as today's gerbils.

Gerbils live in the treeless semiarid deserts of the northern Caucasus and Armenia, and burrow in sands and clays for protection from predators. The fossil multituberculates also seem to have lived in a semiarid, desert environment as they appear to have been buried by wind-blown sand.

There are other similarities, too. The fossil skulls have comparatively large eye sockets, which suggests that, like gerbils, *Nemegtbaatar* and its relatives were big-eyed, nocturnal creatures. The well-preserved pelvis of another Mongolian species, *Kryptobaatar*, also gives a clue as to how the multituberculates reproduced. The narrow canal through which their young would have passed measures just ¼in (3 mm) across. This is smaller than any known shelled or membrane-covered egg, which suggests that the multituberculates gave birth to live young – like living placental mammals – rather than laying eggs. However, the babies would have been extremely small and undeveloped (when they are usually described as neonates) and so more like the young of the marsupials.

NIGHTLIFE *The large eye sockets of this Cretaceous multituberculate mammal from Mongolia indicate big eyes and nocturnal habits.*

Such helpless and vulnerable young would have needed the security of an underground nursery in a burrow to survive.

Living alongside the multituberculates, another group of small mammals had a number of features – such as a first set of 'baby' teeth replaced by new adult ones – that relate them to the dominant living mammals, the placentals. Shrew-like forms such as *Kennalestes*, for example, a primitive insectivore, and the slightly larger *Zalambdalestes*, which has been compared with the living African elephant shrew, nurtured their embryos for longer than the multituberculates, and ultimately went on to inherit the modern world.

THE ADVANTAGE OF ISOLATION

The marsupials are an ancient group whose ancestry stretches way back to Late Cretaceous times, 80 million years ago, in North America. From there, the marsupials spread almost globally, as many of the continents were still connected, and by Tertiary times, 55 million years ago, they were resident in North America, Eurasia and Africa. These marsupials were later replaced by the more modern mammals, but some survived in South America and the Australian continent, where they developed into forms that were remarkably similar to those of their very distant, modern mammal relatives.

When North and South America were reconnected by land about 3 million years ago, the last of several extraordinary fauna exchanges took place. Mastodonts, pumas, bears, camels, horses, dogs and rabbits,

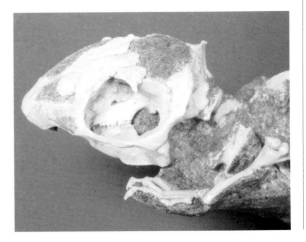

among many others, swept south, while monkeys, porcupines, armadillos, ground sloths, glyptodonts and opossums headed north. Although it was thought that the North American invaders wiped out the species living farther south – especially the marsupials – it now appears that many of the animals in South America happily co-existed until the great extinctions of the Pleistocene era, which drastically affected all the groups of larger mammals.

The South American carnivorous marsupials that developed included various dog-like animals, a sabre-tooth cat – *Thylacosmilus* – and some bear-like creatures. These preyed on plant-eating marsupial 'horses' and 'deer'. Some of the most bizarre mammals ever to have evolved anywhere were part of the South American marsupial group, and included the glyptodont armadillos and the giant ground sloth *Megatherium*.

CROSSING THE BRIDGE

Sloths first appeared in the Early Tertiary era, 40 million years ago, and spread throughout the Americas. Their distribution is proof of a two-way exchange between North and South America whenever a land bridge formed between the two, as there is today. The largest of the sloths to develop was *Megatherium*, a huge, 20 ft (6 m) plant-eater which could rear up on its massive hind legs and pull down high branches with its sickle-shaped claws. They became extinct in the Pleistocene era 11 000 years ago. Most were smaller than this, but the sloths' success as a group meant that they overlapped with modern humans, who may have played a role in their near extinction.

The more prolonged isolation of the Australian continent meant that both primitive monotreme mammals and marsupials were able to survive and flourish without interference from other more advanced placental mammal groups, and, as a result, became remarkably diverse. The monotremes are that puzzling group of animals that combine mammalian features – such as hair, and suckling the young on breast milk – with reptilian-type laying of shelled eggs. Today, they only survive in the Australian continent as the platypus (*Ornithorhynchus*) and the spiny anteater (echidna). Exactly how they are related to other mammal groups is not yet clear, largely because so few monotreme fossils have been found over the years.

One fossil that has been found in Australia is that of a Lower Cretaceous jaw fragment and teeth. Between that find and the fossils of Mid Tertiary (Miocene)

THE HIDDEN TALENTS OF GLYPTODON

In Late Tertiary times, giant armadillos which grew to over 10 ft (3 m) long and 5 ft (1.5 m) high, and weighed up to 2 tons, were common plant-eaters in the savannahs of South America. They were among the most unlikely mammals ever known. *Glyptodon* had an extraordinary body armour, which was a bit like a tortoise shell but was made up of a mosaic of bony plates. The armour of these bizarre creatures weighed more than 440 lb (about 200 kg). In some species, the armoured tail ended in a spiked bony club.

They looked more like armoured ankylosaur dinosaurs such as *Polacanthus* than mammals, and their massive size and weight suggested that they were slow, ponderous animals, whose bony armour was a protection against large marsupial sabre-toothed cats.

But a new analysis of how the glyptodont skeleton worked suggests that they were even odder than previously thought. The study turned up an extraordinary discrepancy between the strength of their massive back legs and their more slender front legs, suggesting that the animals were actually able to stand upright. The thickness of the thigh bones and their enormous muscles provided more than sufficient strength to carry the total body mass alone, which suggests that they regularly walked on two feet. Indeed, if the males had not been able to stand on their hind legs, they would not have been able to reproduce. This may well be the reason that they developed the ability to stand up in the first place.

Many *Glyptodon* remains show healed scars on the carapaces, wounds which may well have resulted from the males fighting over the females. By standing up on their hind legs, the males would have been able to display their mass to maximum effect and bring their weight to bear down on any rival.

ARMOUR PLATED *One of the most extraordinary mammal adaptations is the shell of the giant South American armadillo* Glyptodon.

monotremes, 100 million years later, there is a large gap. Later evidence, from the Pleistocene era, shows that some echidnas evolved to a considerable size. *Zaglossus*, for example, reached a weight of 44 lb (20 kg).

The oldest marsupial fossils from the Australian continent date back to the Early Tertiary (Oligocene) era, 30 million years ago. From then on, as in South America, their development paralleled that of the placental mammals to an astonishing extent. The marsupial, wolf-like animal *Thylacinus* has an incredibly dog-like skull, and there were marsupial 'moles', 'mice', 'cats', 'anteaters' and even 'lions' (*Thylacoleo*).

The famous kangaroos and wombats also developed, with one group of wombats evolving into megaherbivores during the Pleistocene era. These hippopotamus-sized animals (*Diprotodon*) weighed well over 2200 lb (1000 kg) and grazed the open grasslands in great herds.

THE PLACENTAL MAMMALS

The extraordinarily rapid diversification of the mammals, as shown by the fossil records, must be one of the most incredible events in the whole history of life. It seems

PROVING THE THEORY OF EXTINCTION

In the late 18th century, the fossil remains of *Megatherium*, a 12 ft (3.7 m) long animal, were found in Paraguay. Published engravings of the bones were sent to the renowned scientist Baron Georges Cuvier in Paris, who went on to show that not only was *Megatherium* new to science but that it was also extinct.

To prove that this was a different animal to any seen before, he applied his new principles of comparative anatomy. These showed that this rhinoceros-sized mammal had a close relationship to the living sloths, despite the vast difference in size and life habits.

At this stage in the understanding of the history of life, it was not clear whether any organisms had become extinct or not. But Cuvier realised that the best chance of proving extinction lay with the large land animals since they are the most conspicuous. And so he argued, for the first time, that *Megatherium* was sufficiently distinct from any known living creature to warrant recognition

as an extinct animal. Nonetheless, Baron Georges Cuvier was enough of a product of his day to assert that the cause for *Megatherium*'s extinction was Noah's Flood.

SUMO SLOTH *At around 12 ft (3.7 m) in height, Megatherium may have been a gentle giant herbivore but the hands were equipped with wicked-looking claws.*

that within 65 million years, from the beginning of the Tertiary period to the present day, the mammals evolved into nearly 4000 species. Furthermore, much of this diversification was concentrated into about 10 million years, during which 15 or more new groups or lineages occurred. Today, the most familiar groups of mammals that developed are the Carnivora (which includes cats and dogs), Insectivora (including shrews and hedgehogs), Primates (including lemurs, monkeys, apes and humans), Rodentia (including 1700 living species of rats, mice and squirrels), and the Chiroptera

THE LOST SPECIES *In 1936, one of the few remaining marsupial animals became extinct with the death of the last Tasmanian wolf.*

(bats), which have over 1000 living species. A number of much less familiar, and now extinct, groups of mammals also developed. These included the insect-eating leptictids (such as *Leptictis*), the dog-like arctocyonids (such as *Arctocyon*), the pig-like taeniodonts (for example *Stylinodon*), the sheep-sized pantodonts (such as *Titanoides*) and the large group of sheep-sized condylarths (such as *Phenacodus*). Larger predators included the wolf-sized mesonychids (such as *Mesonyx*), the Carnivora-type creodonts (such as *Oxyaena*) and the largest mammals of the time, the dinocerates, which included the rhinoceros-sized *Uintatherium*.

A QUESTION OF SIZE

Today, as throughout the Tertiary era, the herbivore mammals vastly outnumber their predators, the carnivores. Just as the plants they feed on vary in size from ground-hugging mosses and lichens to giant trees soaring up to 300 ft (91 m) into the air, so do the plant-eaters. The smallest and most abundant of the herbivores are the extraordinarily successful and adaptable rodents, while the larger herbivores today include two major groups – the ungulates and the proboscideans. The ungulates are by far the bigger group and include horses and rhinoceros (known as odd-toed perissodactyls), and cattle, deer, pigs, sheep and camels (known as even-toed artiodactyls).

ARMED TO THE TEETH

The success of the rodents is largely due to their highly specialised teeth which – unusually for mammals – grow all the time. Mice, squirrels, beavers and guinea pigs are all capable of chewing their way through tough, woody plant tissue such as the husks of seeds, nuts and roots, which gives them a distinct advantage over herbivores with weaker teeth. Cutting through the tough protective shells and husks of this type of plant food quickly wears their teeth down, however, which is why their chisel-shaped incisors are continually growing.

Most of the earliest rodents from the Early Tertiary era of North America and Eurasia, about 40 million years ago, were small creatures much like modern rats and mice, although one group, the caviomorphs of South America, were ancestors of today's guinea pigs, chinchillas and capybara. Some of these weighed more than 100 lb (45 kg), but that is nothing compared with the capybara's giant ancestor, *Telicomys*. This unspeakably huge rodent, which developed in the Late Miocene era, about 12 million years ago, was the size of a rhinoceros.

HORSES ON THE RUN

So many horse fossils have been found in the terrestrial sediments of North America and Eurasia that it has been relatively easy

HORSES FOR COURSES   In less than 55 million years, horses evolved from the dog-sized Hyracotherium *(above) to the modern* Equus idahoensis *(above right).*

TOEING THE LINE   The increase in horses' body size and the reduction in the number of toes – from four toes on the hind legs to a single toe – offer a classic example of Darwinian theory in action.

to trace equine evolution over the last 55 million years. Fossil evidence shows that, as time progressed, each successive version of the modern-day horse was a larger, faster version of its predecessor.

Its improved running ability was created by a gradual reduction in the number of its toes. The earliest, the dog-sized *Hyracotherium* of Eocene times, had four toes on its front feet, while *Mesohippus* of the Oligocene era had just three. In the Pliocene era, the one-toed *Pliohippus* developed, as did the modern *Equus*. These changes over 55 million years have been accompanied by a move from browsing on leaves to grazing on grass, which led to modification of the horse's teeth and jaw muscles.

Less is known about the development of the living rhinoceros, the last survivor of an abundant and diverse group that once included by far the largest of modern land animals. *Indricotherium*, from the Oligocene period, was the only mammal that could in any way compete with the sauropod dinosaurs in terms of size and bulk. It stood 16 ft (5 m) high at the shoulder and weighed perhaps as much as 30 tons. Such bulk can generally be sustained only by a plant-eating habit. Nevertheless, the creature must have consumed prodigious quantities of plant food each day. As a

TWO-PRONGED DEFENCE
Arsinoitherium was a rhino-like browsing herbivore that lived in Africa and southern Europe 35 million years ago.

HOOF AND CLAW *The herbivore Chalicotherium could probably walk like an ape on its hind legs and occasionally rest on the knuckles of its long arms.*

comparison, an elephant – which weighs only 5 tons or so – requires up to 440 lb (200 kg) of food a day. Horned rhinoceros ranged widely throughout the world during the Mid Tertiary era and adapted to a variety of different climatic conditions. *Coelodonta*, the woolly rhinoceros of Eurasia, was even able to survive the glacial conditions of the last Ice Age.

EXTINCT KNUCKLE-WALKERS

Two perissodactyl groups which are now completely extinct are the brontotheres and chalicotheres, which included some large, unusual looking animals. The brontotheres were rhinoceros-like, while the chalicotheres were very different indeed. Their extraordinarily long arms ended in three fingers bearing hoof-like nails, while their shorter hind limbs had small claws. It is thought that *Chalicotherium*, one of these peculiar animals, stood about 8 ft (2.4 m) at the shoulder and may have engaged in a form of knuckle-walking which is normally only associated with gorillas and chimps.

The chalicotheres were

CUDS AND A GOOD DIGESTION
South American llamas were some of the first herbivores capable of digesting tough grasses by chewing the cud.

the cheek teeth distinctive of all cattle, deer, sheep and camels today. As ruminants, they used their cheek teeth to help process their food, regurgitating it or chewing the cud to extract the maximum food value from the plant material. The grinding surfaces of the teeth were like files, with crescent-shaped ridges, and the lower jaw was capable of the rotary motion so characteristic of domestic cattle and sheep.

The closest living relatives of the oreodonts are camels and llamas, which also started from relatively small animals such as the goat-sized *Poebrotherium* (found in the early Oligocene era of North America). They spread to South America, Asia and Africa in Miocene times, and later adapted easily to domestication – one of the reasons cited for their success. By comparison, the Carnivora have been relatively less successful, with only cats and dogs adapting to domestication on any large scale.

PIGS AND HIPPOS

Another successful group of even-toed hoofed mammals are the bunodonts, which include the pig and the hippopotamus. Once they had originated in North America, the early bunodonts quickly became widespread – as their abundance as fossils in Africa shows. This has led to them being used to distinguish successive subdivisions of rock layers, and so to accurately date some of the earliest of fossil hominids found in East Africa.

In the past, hippos were much more widely distributed than they are today, moving as far north as the British Isles. There, they lived alongside elephants, giant deer, bison, hyenas, lions and wolves during the warm, interglacial phases of the last Ice Age. Their success lay in part in their teeth, which were designed for eating aquatic plants. Modern hippos, one of the most recently evolved mammal groups, arose in the
continued on page 146

herbivores, and probably used their hooves and claws for digging up plant roots.

OF CAMELS, CATTLE AND SHEEP

The artiodactyls are a large and diverse group of herbivores which are distinguished by having an even number of toes.

Their third and fourth toes are equally enlarged and bear most of the weight of the body. Within this group are the oreodonts, creatures which started out the size of rabbits but by Oligocene times, 30 million years ago, had grown to a similar stature as a modern pig. By then they had evolved

OLIGOCENE WORLD   *Around 30 million years ago, the Earth's landscapes finally took on a modern aspect. Songbirds filled the air with sound, and flowering plants – such as newly evolved grasses, herbs, shrubs and trees – provided a blanket of colour. Most important were the grasses, which invaded the open semiarid savannah and provided food and cover for a range of animals – from rodents to ancestors of the horses (Mesohippus) to extinct giants such as Brontotherium. The plant-eaters became prey for ancestors of carnivores such as the hyena-like Hyaenodon and the dog-like Hesperocyon.*

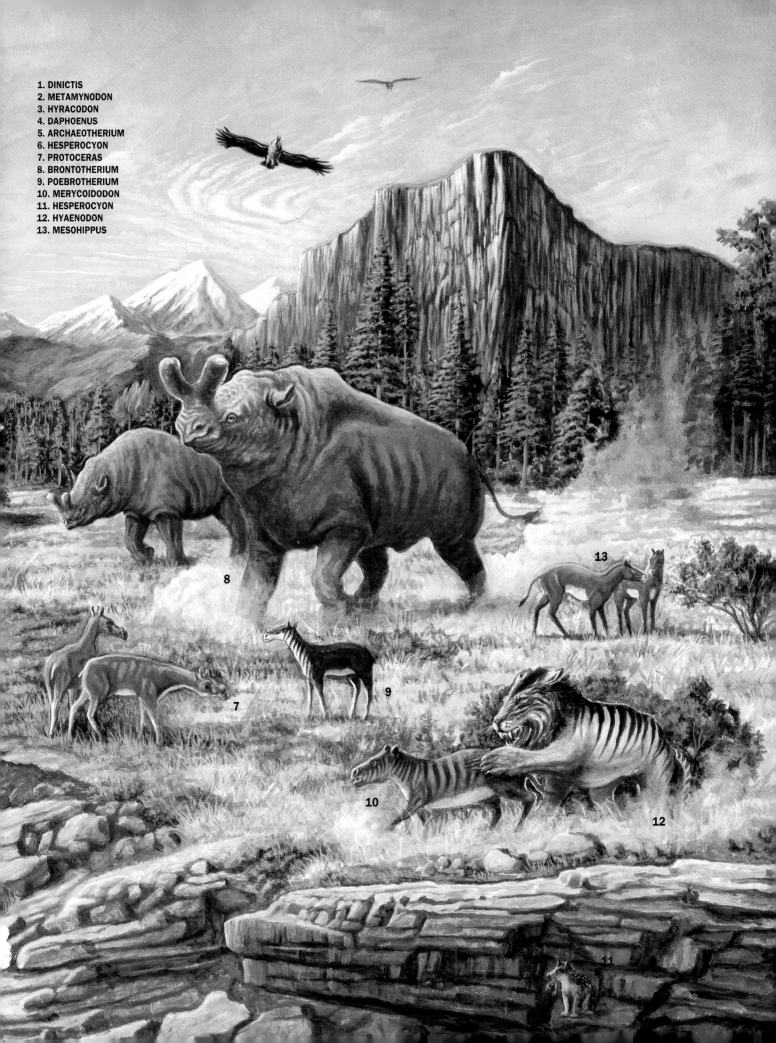

1. DINICTIS
2. METAMYNODON
3. HYRACODON
4. DAPHOENUS
5. ARCHAEOTHERIUM
6. HESPEROCYON
7. PROTOCERAS
8. BRONTOTHERIUM
9. POEBROTHERIUM
10. MERYCOIDODON
11. HESPEROCYON
12. HYAENODON
13. MESOHIPPUS

STEGODON

DEINOTHERIUM

GOMOPHOTHERIUM

ANCIENT TUSKERS *From pig-like animals, the proboscideans have evolved trunks and a variety of tusk forms over 50 million years.*

Pliocene era from ancestors such as *Bothriodon*, which were widespread in Africa, North America and Eurasia. They had a similar lifestyle to the modern hippopotamus, which replaced them when they died out.

TAKING TO TUSKS

The familiar but dwindling herds of the only two remaining species of trunk-bearing proboscideans – the Asian and African elephants – are a reminder of yet another group of megaherbivores that were much more numerous and widespread in the past. Once, more than 160 proboscidean species roamed the Earth's landscapes in vast numbers, but today the nearest living elephant relatives – sirenids, sea cows,

AMBELODON

dugongs and manatees – are continually threatened with extinction. Yet again, according to fossil records, this important group appears to originate from the Early Eocene, 50 million years ago.

Moeritherium was 3 ft (1 m) long, an amphibious sirenid, or hippo-like animal, which lived in North Africa on the margins of the Tethys sea. Most modern

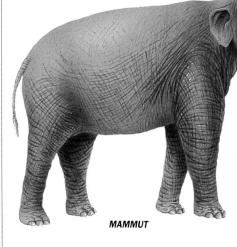

MAMMUT

proboscideans show specialised characteristics adapted to their surroundings, but *Moeritherium* does not, although it did have six pairs of cheek teeth which operated together like those of modern mammals.

The most typical characteristics of the proboscideans, however, are a trunk and tusks. The earliest proboscideans to show the beginnings of these features were the deinotheres, one of which – *Deinotherium* – appeared in the Miocene era 25 million years ago. The presence of a distinctive keyhole-shaped nasal opening in the skull suggests that the

animal had a trunk that was formed by a fusion of the nose and upper lip. Its lower incisors were modified to form tusks, which curved downwards, and may have been used for digging. Some extinct proboscideans, such as *Gomphotherium*, had upper and lower pairs of tusks, but the pattern adopted by the mastodonts and elephants – where the tusks only developed from the upper incisors – was more common.

The proboscideans flourished during the Miocene era up until around 5 million years ago, and the mastodons – as the Miocene forms are collectively known – included several

different families. The mastodon *Mammut* existed in the Americas as recently as 10 000 years ago, while another family, the stegodons, survived into the Ice Age (50 000 years ago) in South-east Asia and Africa.

The stegodons, including *Stegodon* itself, were the first proboscideans to develop the ridged, grinding molar teeth that characterise true elephants. The six pairs of cheek teeth appeared successively, each set larger than those before. The sixth and last pair would appear when the animal was about 30 years old. Each tooth was more than 1 ft (30 cm) long and weighed 4 lb (1.8 kg).

To deal with the fibrous and often woody nature of its plant food, *Stegodon's* molars had a series of hard enamel ridges which acted as rasping files. The teeth of the upper and lower jaw ground the plant material between them, a technique which

successfully broke down the plants for digestion but led to intense wear on the teeth themselves. By an old age of 70 or so, the sixth and last pair of molars would have been almost completely ground away.

About 5 million years ago, the elephants split into three main branches: the Asian elephant (*Elephas*), the African elephant (*Loxodonta*) and the mammoths (*Mammuthus*). The first mammoths evolved as woodland browsers in tropical Africa, from where they spread into Europe, Asia and eventually North America.

As the Northern Hemisphere moved into the Ice Age, mammoths were able to adapt to the increasingly cold climate. They evolved into magnificent woolly creatures such as the *Mammuthus primigenius*, which stood up to 11 ft (3.4 m) high and weighed up to 6 tons. It had a long-haired coat, a protective fat layer and was armed with huge tusks as much as 12 ft (3.7 m) long. The mammoth's distant relatives, the

COOL FOR CATS Big cats such as this Siberian tiger (Panthera tigris altai) were adapted to the cold, and were common in the Ice Age.

elephants, were not as able to adapt to extremes of cold, but at various times of changing climate within the Ice Age the two did coexist. As the climate improved towards the end of the Ice Age, it was the elephants rather than the mammoths that managed to survive (although, some dwarf mammoths survived until just 4000 years ago – around the time the pyramids were being built in Egypt).

STRATEGIES FOR SURVIVAL

For all life forms, there is a range of survival strategies to choose from. On the one hand, there is the policy of small size and abundance – the strategies employed by life forms as diverse as grasses, insects and rodents. On the other, there is gigantism – where size is more important than numbers, speed or guile. The giant redwoods, elephants and whales have all survived over a long period of time using this strategy.

For plants and plant eaters, gigantism is a good option, providing almost unassailable

continued on page 150

PRESERVATION IN ICE

Thanks to the Ice Age, which preserved mammoth remains in deep-freeze like conditions, the anatomy of this extinct creature is known in detail. But while rapid dehydration in subzero temperatures and entombment in ice create the ideal conditions for preservation, it is not true fossilisation. Even though the permafrost Ice Age cadavers of mammoth, woolly rhino, bison and horse from Siberia and Alaska are tens of thousands of years old, the ice which is keeping the animals so well preserved will melt, eventually.

In the meantime, however, they have provided some of the most fascinating and evocative finds of the century. The first was the frozen Beresovka mammoth from Siberia, found in 1901, but perhaps the most remarkable were dug out of the ice in the 1970s and 1980s. In 1977, a complete 40 000-year-old baby male mammoth aged 6-12 months was discovered, and nicknamed Dima; while in 1988 an even younger female, Mascha, was found. Together, they provided much information about the detailed anatomy and diet of mammoths, and even yielded up samples of mammoth DNA. This was first extracted in 1994 and shows some striking differences to the genetic material of living elephants.

THE BIG FREEZE In 1901, it took ten months and a 6000 mile (9600 km) journey to bring this 30 000-year-old mammoth back to St Petersburg in Russia.

EVE'S MAMMOTH

From Condover, a small farming village in the border country between England and Wales, you can see the grey Welsh mountains away to the west across Shropshire's lush fields and pastures. The scene was very different 15 000 years ago, however. Then, the enormous ice sheets of the last glacial cold phase still poured out of the Welsh mountains towards Shropshire, although they were, at last, beginning to melt and retreat.

As the climate became warmer, multi-channelled streams carried the cold and murky meltwaters away from the glaciers, the water forcing its way through all the muck and rubble dumped by the glaciers as they melted. The streams merged together, sorted and washed the rubble and carried it away downstream, only to redeposit it wherever the current slowed. In spring and summer, the local rivers became torrential floods capable of carrying away anything

that was movable, their power and size enormously greater than any of the rivers and brooks of the area today. It must have been around this time that the bones discovered by Eve Roberts were laid down.

IDENTIFYING THE UNEXPECTED

One Saturday afternoon in late September, 1986, Eve's walk took her past an open sand and gravel quarry. As she looked down at the quarry workings, her eyes were drawn to a large brown object sticking out of the muck. Looking closer, she thought that it might be a bone, although this seemed absurd as it was too big to have belonged to a cow. Nevertheless, she phoned the local museum when she returned home.

Jeff McCabe of the Shropshire Museum knew that this area of Britain is a veritable goldmine for prehistory. Almost anything – from 17th-century musket balls to 400-million-year-old fossil trilobites – could turn

up, and Shropshire is the right kind of place for an Ice Age find. He decided to take a look for himself.

Looking around the site in the failing light, he and Eve not only found the unusually large bone that she had seen earlier, but they also spotted what looked like a giant thigh bone 4 ft (1.2 m) long. Clearly this was no cow, but the bones looked remarkably fresh, quite unlike most fossils.

McCabe contacted Dr Russell Coope of Birmingham University, an international expert on Ice Age climate and environments. He in turn brought in a young scientist at Cambridge University, Adrian Lister, who was fast making his name for his research on the large animals – known as the megafauna – of the time. They needed no more than a single look to confirm that this was potentially a big and interesting find. The major problem was that the area where the bones were found had been bulldozed

recently – which meant that they had originally come from somewhere else in the quarry. The only chance of finding the rest of the remains was for the quarry workers to make a giant heap of all the mud that had originally contained the bones and leave the scientists to sort through it.

THE SHROPSHIRE VOLUNTEERS

Raking through the hundreds of tons of bulldozed earth was an enormous task that could not be done alone. Soon, an army of volunteers became involved, forming a type of human conveyor to remove and sort mud by the bucketload. Any bones that were found had to be cleaned in buckets of water with stiff bristle brushes to remove the black sticky mud. Then, they were carefully examined for cracks or other signs of deterioration, as drying out can do more damage to fossils in a few hours than occurred during tens of thousands of years of burial in the ground. Any weaknesses have to be carefully treated with the appropriate chemicals in a suitable environment.

Although progress was slow, a significant proportion of the mammoth skeleton was recovered – including a number of huge leg bones. Then, to everyone's delight, a small but entire lower jawbone was found, complete with its distinctive molar grinding teeth, clearly implying that there was at least one baby mammoth to find in all the debris.

How and why the mammoths had died was explained by examination of the quarry site, which showed that at the end of the Ice Age it had contained a 'kettle hole'. This was a deep natural hole in the surrounding glacial deposits, formed when a large block of buried ice eventually melted. The remaining crater was more than 30 ft (9 m) deep and soon filled with water, surrounded on its edges with marsh and pond plants. Any plant-eater seduced by such succulent grazing could easily fall in and drown. The mud and cold water made excellent natural preservatives for bone by excluding oxygen and so slowing down decay.

THE BONE JIGSAW PUZZLE

As time wore on, the bone count became increasingly impressive – surprisingly few bones had been broken by the bulldozers, and most that had were recovered and stuck back together. What had not been found, however, was the skull and ivory tusks. A mammoth's tusks can grow up to 12 ft (3.7 m) long and so should have been easily recognised, but the huge skull and heavy tusks can become detached from the rest of the corpse and end up being buried in completely different locations.

Eventually, the remarkable total of 400 bones were recovered. When finally assembled, neither the Condover mammoth's skull nor its tusks were found, but most of the rest of the adult skeleton was, plus part of the remains of at least three babies aged between three and six years old. The adult was a mature, 28-year-old male that seemed to be in good health, as far as could be told from the bones, although he had fractured a shoulder blade quite badly. His age was calculated by looking at his teeth. His sixth set of molars were just beginning to replace the worn down fifth ones when he died.

Radiocarbon dating finally established when exactly the mammoths were roaming the Welsh borderlands. It seems they lived between 12 700 and 12 300 years ago, which makes them by far the 'youngest' mammoths ever found in Britain, and provides the first evidence that British mammoths survived the last retreat of the ice.

WATERY GRAVES *Among the most lethal hazards facing mammoths were steep, slippery-sided and water-filled 'kettle holes'.*

safety as long as they reach adulthood. But for the predator who needs speed to catch its prey, too much weight can be a disadvantage. For land animals, for example, the 794 lb (360 kg) weight of an adult male Siberian tiger is about as heavy as an animal

CARNASSIAL CHEEK *With fangs for stabbing and holding, and sharp scissor-like cheek teeth for cutting, the cheetah is a typical carnivore.*

can get before it starts to lose speed. The largest predatory mammal known to have existed was a Miocene hyaenodont, *Megistotherium*. It weighed probably 1764 lb (800 kg), and had long canines rather like those of a sabre-tooth tiger. How fast it could run is unknown, but it must have had enough speed to catch its prey.

THE MEAT-EATERS

Some big, some small, some fierce, some friendly – the Carnivora make up some 236 species of diverse creatures. They represent one of the largest and most important of mammalian groups, with only the 374 species of insectivores,

the 981 species of bats and 1729 types of rodent forming larger groups. It is an extraordinary fact that, within a geologically short time span, the Carnivora 'exploded' from an unpromising beginning of squirrel-sized, insect-eating ancestors into the whole living gang of what has been evocatively called the 'velvet claws'.

Sabre-tooth tigers, modern domestic cats, polar bears, pandas, wolves, whippets, walruses and weasels all come under the same Carnivora classification, and many, but not all, are meat-eating. One of the better known exceptions is the giant panda, whose preferred diet is bamboo shoots.

The one distinguishing characteristic of the carnivores is a particular kind of cheek tooth called the carnassial. These teeth are arranged like shears, with opposing pairs in the upper and lower jaws. This allows the animal to cut the flesh into small pieces more effectively, which helps speedy consumption and digestion. Sharp, stabbing fang-like canine teeth are also often seen as a common characteristic, although not all carnivores possess them.

The best way to understand how the various extinct Carnivora lived is by comparing them with living carnivores, or by piecing together what the animal looked like and how it functioned. There are rules that link behaviour with ecological circumstance and feeding habit – no animals live in isolation from others or their habitat, nor have they

done so in the past. Fortunately for palaeontologists, the feeding requirements and behaviour of the Carnivora are generally reflected in their bones and teeth, which are often the only part of an animal to be preserved as a fossil. Isolated teeth – whether fossil or modern – can be read almost like a menu.

Among the extinct carnivorous mammals living during the Early Tertiary era, 60 million years ago, were the mesonychids and creodonts. Mesonychids such as *Mesonyx*, a dog-sized animal, had some characteristics of the carnivores – for example, plant-eating molar teeth which had been modified for meat-eating. But they also had hooves on their toes like the horses and yet also showed similarities of skull structure to the early whales.

Andrewsarchus, a Late Eocene mesonychid from the Gobi desert region of Mongolia, had a skull larger than any other known land-living carnivore (it measured 33 in/84 cm lengthways and 22 in/56 cm across). Altogether it was more than 16 ft (4.9 m) long but, again, it had hooved toes, which means that it would not have been capable of grasping its prey. This suggests

LEADERS OF THE PACK *At 16 ft (4.9 m) long, the giant hyena-like* Andrewsarchus *was the largest land-living carnivore ever known.*

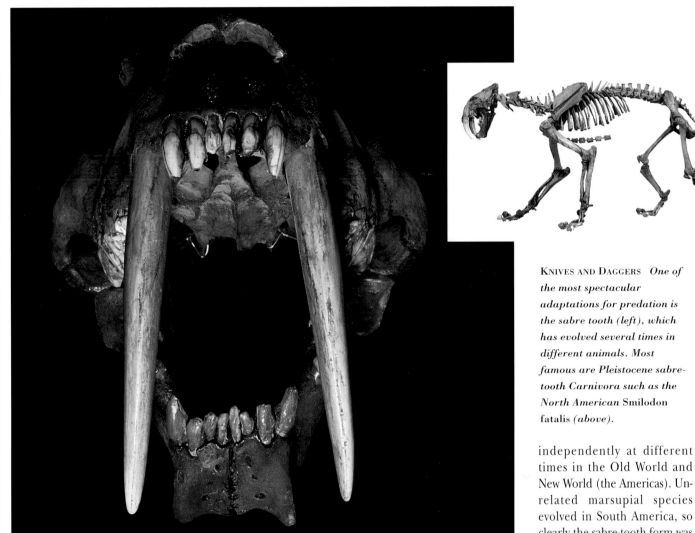

KNIVES AND DAGGERS *One of the most spectacular adaptations for predation is the sabre tooth (left), which has evolved several times in different animals. Most famous are Pleistocene sabre-tooth Carnivora such as the North American* Smilodon fatalis *(above).*

that the animal was probably a scavenger that operated in packs, rather than an active killer.

The creodonts are more likely candidates as the ancestors to the true Carnivora. They were well established throughout North America, Africa and Eurasia by the Early Tertiary era and, for 20 million years, various dog, cat, bear and hyena-like creodonts flourished as carnivores. However, they had not yet acquired the true carnivores' tearing and cutting up of their prey.

Hyaenodon was a hyena-like creodont which had slim legs and feet raised up onto the toes, just as modern dogs do. This suggests that *Hyaenodon* had a similar mode of life to modern hyenas. The hyaenodonts were a particularly successful group of creodonts and eventually produced *Megistotherium*, the biggest creodont and predatory mammal ever known. Its skull measured 26 in (66 cm) long, twice the size

of a tiger's. But the first true Carnivora – small, unspectacular animals – appeared in North America about 55 million years ago. One group, the miacids, were treetop hunters similar to living pine martens. They soon diverged into the viverrines and vulpavines. The viverrines formed the cat branch and, from the Oligocene era onwards, they gave rise to the civets, mongooses, true hyenas and finally cats of the 'Old World' – the interconnected land areas of Europe, Africa and Asia. The vulpavines formed the dog branch, evolving into the bears and then the wolves and foxes.

SABRE-TOOTHED CATS

Perhaps the most famous of the extinct Carnivora are the sabre-toothed cats, which were the dominant large cats at the end of the Miocene era 6 million years ago. These were not a single group, as a number of species of sabre-toothed cats evolved

independently at different times in the Old World and New World (the Americas). Unrelated marsupial species evolved in South America, so clearly the sabre-tooth form was highly successful. All such animals shared a special adaptation of their jaw which allowed the lower half to drop right down, giving an enormous gape to the mouth. The canine sabre teeth of the North American *Smilodon*, for example, were curved blades up to 7 in (18 cm) long, and these had to be accommodated when the mouth opened. Whether these teeth were used for stabbing or cutting is unknown, but the cats were probably ambush hunters that seized their prey by the neck. The tooth blades would have been used to cut into the soft neck tissues near the major blood vessels as the prey was throttled or bled to death.

THE LAST GREAT EXTINCTION

Over the last 2 million years, the evolutionary history of the mammal groups has undergone drastic changes. And the most significant of these has been the wave of extinction that reduced mammalian diversity

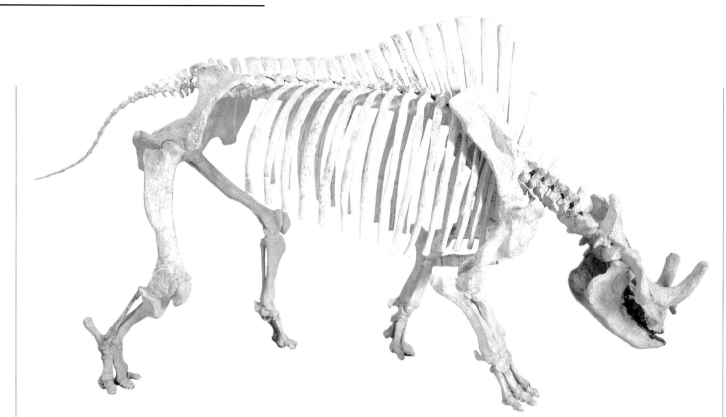

HORNS FOR HEADS *Head gear reflected the sex differences of rhino-like titanotheres, such as this double-horned male* Megacerops aoer.

worldwide. Gone are the mammoth, woolly rhinoceros, giant deer, cave bears, sabre-toothed cats, marsupial carnivores, glyptodonts and giant ground sloths. All of the continents were affected in the wave of extinction, but especially Europe, parts of Asia and North America. Within a period of just 2000 years, somewhere between 12000 and 10000 years ago, North America lost around 73 per cent of its large mammals in 33 different species, ranging from mammoths to deer. South America lost 80 per cent of its mammals, from camels to mastodons, making a total of 46 species in all. In Australia, over 40 species became extinct. These included the diprotodonts, marsupial carnivores and giant wombats. The question as to why this happened has concerned palaeontologists for years. The most obvious cause would seem to be environmental change following the end of the last glacial phase, 10000 years ago. However, the end of previous glacial phases within the Ice Age does not appear to have had such a drastic effect.

One of the explanations given for the extinctions focuses on the fact that many of the animals that disappeared were relatively large. The animals thought to be most at risk to environmental change are the very large herbivores which, as plant eaters, tend to be highly specialised to particular kinds of vegetation. During much of the Ice Age, there were significant changes in the plant cover over much of the planet. There was a dwindling of the vast steppe grasslands in North America and Asia, which

BIGGEST FAN *Spanning 12 ft (3.7 m), the antlers of the Ice Age* Megaloceros *are the largest known. They were shed and regrown annually.*

MAMMOTH COUSINS *The American contemporary of the Eurasian woolly mammoth was Mammuthus colombi (right and re-created in a painting above). This species ranged into more southern, warmer areas such as California, and was bigger and less hairy than its Eurasian 'cousin'.*

had provided feeding grounds for enormous migratory herds of the extremely large herbivores. The grasslands that were left may well have been overgrazed by the hungry hordes of animals, creating a spiral of decline. As the very big herbivores diminished, so did their predators – the top carnivores.

LEARNING FROM THE MAMMOTHS

Some clues as to what happened leading up to the extinction can be found in the history of the mammoths. Just 20 000 years ago, mammoths formed vast migratory herds throughout Europe and Asia, from Northern Ireland to China. By 12 000 years ago, however, they had disappeared from north-western Europe; by 11 000 years ago, they had gone from North America; and by 10 000 years ago, from Siberia.

A dwindling and isolated colony survived on the bleak Wrangel Island in the Arctic Ocean until 3700 years ago, where they gradually became dwarfed. Reducing

food supplies in these inhospitable frozen wastelands caused these vast, massive animals to finally shrink to a height of under 6 ft (1.8 m).

Another possible cause of the demise of the mammoths pinpoints the advent and spread of man, from the early hominids through to the modern humans. Certainly, the beautiful and evocative portraits of mammoths found in limestone caves at Rouffignac in the Dordogne, France, confirm that our Ice Age ancestors were keen mammoth spotters. These Stone Age people could reproduce every detail of the mammoth form with an accuracy that could only have come from close observation of the animal's anatomy and behaviour.

They risked their lives hunting the powerful beasts with primitive weapons – some of the cave paintings illustrate mammoth

hunts – and many Ice Age artefacts were made from mammoth bone and tusks. Mammoth remains excavated at Dent, Colorado, in 1932 provided the first unequivocal evidence that mammoths were killed by early humans. A stone spear point was found wedged in among 11 000-year-old mammoth bones. Other evidence comes from the Ukraine, where over 70 Neanderthal huts made from mammoth bones have been discovered. Although the huts contain the remains of over 100 animals, suggesting wholesale slaughter, dating of the bones has shown that they range over

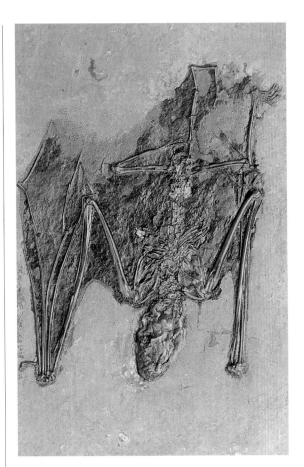

ONE OF THE FEW *Fossil bats, such as the 50-million-year-old* Palaeochiropteryx, *are rare by comparison with their living descendants.*

8000 years in age. Consequently, they must have been collected slowly over a period of time, quite possibly from the remains of animals that died from natural causes.

Whether the mammoth extinction was due to interference from early humans or was the result of environmental changes remains in the realms of the unknown, but comparisons with modern-day elephants are almost inevitable. The preserved stomach contents of frozen specimens from Siberia show that they fed mainly on huge quantities of grasses – one stomach contained around 640 lb (290 kg) of partly digested plants – and, like modern-day elephants, mammoths had to spend up to 20 hours a day eating in order to survive. As the prairie grasslands of Eurasia and North America greatly diminished, so did the mammoth herds – their populations becoming increasingly fragmented and isolated in much the same way as is happening to the elephants today. Research has shown that when the population of a large herd

drops below 1000 or so it will eventually become extinct, so the disappearance of the mammoth and other mammals of the Ice Age does sound a warning note about the continuing survival of remaining large species in the world today.

Parallels between what happened to the mammoths and the predicament of modern elephants are especially hard to avoid as there seems to be a direct correlation between the timing of the arrival of the first human hunters and the decline of the large mammals in all major continents, from North America to Australia.

THE END OF ONE ERA – AND THE START OF ANOTHER

Not all mammal species have suffered to the same extent since the environmental changes of the Ice Age. Smaller forms, such as bats, rabbits, insectivores and the hugely successful rodents, were largely unaffected – as were the Primates.

Comprising 200 species, Primates – which include the modern human *Homo sapiens* – probably originated way back in the Late Cretaceous era, but the first convincing fossil evidence comes from Eocene times, 40 million years ago, with animals such as *Tetonius*. With their grasping hands and feet, long tails, short snouts and large forward-facing eyes capable of stereovision, it is clear these lemur-like animals were

RARE HARE *Entire skeletons of fossil rabbits, such as this* Prolagus, *are extremely unusual. Teeth and bone fragments are more common.*

well adapted for climbing. Needless to say, the story of their subsequent evolution is of great interest to us humans – it is our story too. What is at first surprising is that it should be so contentious and complicated.

Dealing with relatively recent time, there should be plenty of fossil evidence but the habits and habitats of early Primates have ensured that this is not the case. Terrestrial organisms need to be buried rapidly to have much chance of being preserved. But in the economy of nature, the bodies of small, arboreal mammals are effectively 'recycled' through the food chain. Thanks to a variety of carrion feeders and scavengers – from hyena and vultures to insects, worms and bacteria – often the only remains are the indigestible ones, the teeth. Even these may become degraded chemically.

By comparison with the fossil record of the elephants, pigs or rodents, that of our human ancestors is exceedingly fragmentary and problematic. With its complexity and difficulty of interpretation, human evolution is a story in its own right.

One particularly evocative fossil marks the end of the global domination of the very large – or mega – mammals. In 1979, Kenyan palaeontologist Mary Leakey and her coworkers uncovered a layer of hardened mud, which had been baked by a fall of hot ash from a volcanic eruption in the continuously active Rift Valley in Tanzania. Because of the association with volcanic rocks, the layer can be dated to about 3.6 million years ago. The footprints included numerous tracks of deer and pigs crisscrossing the soft mud surface. But among all the different prints made by animals are

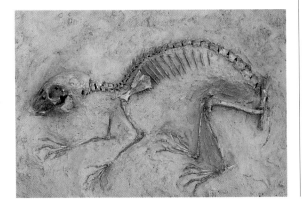

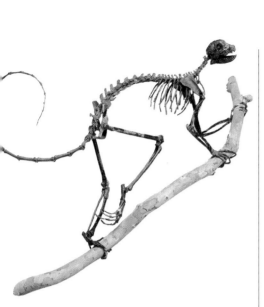

FOOTPRINT EVIDENCE *The 'Laetoli pavement' (right) was proof positive that hominids were walking upright by 3.6 million years ago. Long-tailed, lemur-like Primates, such as Smilodectes (above) from Wyoming, may have been ancestors of man.*

three sets of unmistakable human-type footprints, heading in a straight line at a steady walking pace. The size difference suggests two adults and a child.

This is the first unequivocal evidence of hominids walking upright (with an 'upright bipedal stance' as it is technically called). The similar quality of the prints, their closeness and parallelism shows that they were made at the same time. We do not know the sex, we do not know where they were going or whether they were holding hands. Very little is known about them at all, which is perhaps why they are so evocative – viewers today can come up with their own interpretation to fit the evidence.

None of the higher apes, such as chimps, that can walk bipedally today could produce prints quite like this. If walking upright in this way is one of the critical human attributes then man is at least 3.6 million years old, with ancestors in Africa. Not since Neil Armstrong took the first walk in lunar territory on July 20, 1969, has a set of footprints signified so much.

PICTURE CREDITS